# Tensor Analysis and Continuum Mechanics

Wilhelm Flügge

Springer-Verlag New York Heidelberg Berlin 1972

Dr.-Ing. Wilhelm Flügge
Professor of Applied Mechanics, emeritus
Stanford University



With 58 Figures

ISBN 0-387-05697-1 Springer-Verlag New York Heidelberg Berlin
ISBN 3-540-05697-1 Springer-Verlag Berlin Heidelberg New York

# Preface

Through several centuries there has been a lively interaction between mathematics and mechanics. On the one side, mechanics has used mathematics to formulate the basic laws and to apply them to a host of problems that call for the quantitative prediction of the consequences of some action. On the other side, the needs of mechanics have stimulated the development of mathematical concepts. Differential calculus grew out of the needs of Newtonian dynamics; vector algebra was developed as a means to describe force systems; vector analysis, to study velocity fields and force fields; and the calculus of variations has evolved from the energy principles of mechanics.

In recent times the theory of tensors has attracted the attention of the mechanics people. Its very name indicates its origin in the theory of elasticity. For a long time little use has been made of it in this area, but in the last decade its usefulness in the mechanics of continuous media has been widely recognized. While the undergraduate textbook literature in this country was becoming "vectorized" (lagging almost half a century behind the development in Europe), books dealing with various aspects of continuum mechanics took to tensors like fish to water. Since many authors were not sure whether their readers were sufficiently familiar with tensors, they either added a chapter on tensors or wrote a separate book on the subject. Tensor analysis has undergone notable changes in this process, especially in notations and nomenclature, but also in a shift of emphasis and in the establishment of a cross connection to the Gibbs type of vector analysis (the "boldface vectors").

Many of the recent books on continuum mechanics are only "tensorized" to the extent that they use cartesian tensor notation as a convenient shorthand for writing equations. This is a rather harmless use of tensors. The general, noncartesian tensor is a much sharper thinking tool and, like other sharp tools, can be very beneficial and very dangerous, depending on how it is used. Much nonsense can be hidden behind a cloud of tensor symbols and much light can be shed upon a difficult subject. The more thoroughly the new generation of engineers learns to understand and to use tensors, the more useful they will be.

This book has been written with the intent to promote such understanding. It has grown out of a graduate course that teaches tensor analysis against the background of its application in mechanics. As soon as each mathematical concept has been developed, it is interpreted in mechanical terms and its use in continuum mechanics is shown. Thus, chapters on mathematics and on mechanics alternate, and it is hoped that this will bring lofty theory down to earth and help the engineer to understand the creations of abstract thinking in terms of familiar objects.

Mastery of a mathematical tool cannot be acquired by just reading about it—it needs practice. In order that the reader may get started on his way to practice, problems have been attached to most chapters. The reader is encouraged to solve them and then to proceed further, and to apply what he has learned to his own problems. This is what the author did when, several decades ago, he was first confronted with the need of penetrating the thicket of tensor books of that era.

The author wishes to express his thanks to Dr. William Prager for critically reading the manuscript, and to Dr. Tsuneyoshi Nakamura, who persuaded him to give a series of lectures at Kyoto University. The preparation of these lectures on general shell theory gave the final push toward starting work on this book.

Stanford, California W. F.

# Contents

CHAPTER 1

# Vectors and Tensors

It is assumed that the reader is familiar with the representation of vectors by arrows, with their addition and their resolution into components, i.e. with the vector parallelogram and its extension to three dimensions. We also assume familiarity with the dot product and later (p. 36) with the cross product. Vectors subjected to this special kind of algebra will be called Gibbs type vectors and will be denoted by boldface letters.

In this and the following sections the reader will learn a completely different means of describing the same physical quantities, called tensor algebra. Each of the two competing formulations has its advantages and its drawbacks. The Gibbs form of vector algebra is independent of a coordinate system, appeals strongly to visualization and leads easily into graphical methods, while tensor algebra is tied to coordinates, is abstract and very formal. This puts the tensor formulation of physical problems at a clear disadvantage as long as one deals with simple objects, but makes it a powerful tool in situations too complicated to permit visualization. The Gibbs formalism can be extended to physical quantities more complicated than a vector (moments of inertia, stress, strain), but this extension is rather cumbersome and rarely used. On the other hand, in tensor algebra the vector appears as a special case of a more general concept, which includes stress and inertia tensors but is easily extended beyond them.

## 1.1. Dot Product, Vector Components

In a cartesian coordinate system $x$, $y$, $z$ (Figure 1.1) we define a reference frame of unit vectors $\mathbf{i}_x$, $\mathbf{i}_y$, $\mathbf{i}_z$ along the coordinate axes and with their help a force vector

$$\mathbf{P} = P_x\mathbf{i}_x + P_y\mathbf{i}_y + P_z\mathbf{i}_z \tag{1.1a}$$

and a displacement vector

$$\mathbf{u} = u_x\mathbf{i}_x + u_y\mathbf{i}_y + u_z\mathbf{i}_z. \tag{1.1b}$$

These formulas include the well-known definition of the addition of vectors by the parallelogram rule.

In mechanics the work $W$ done by the force $\mathbf{P}$ during a displacement $\mathbf{u}$ is defined as the product of the absolute values $P$ and $u$ of the two vectors and of the cosine of the angle $\beta$ between them:

$$W = Pu \cos \beta.$$

This may be interpreted as the product of the force and the projection of $\mathbf{u}$ on the direction of $\mathbf{P}$ or as the product of the displacement and the projection of the force on $\mathbf{u}$. It is commonly written as the dot product of the two vectors:

$$W = \mathbf{P} \cdot \mathbf{u} = \mathbf{u} \cdot \mathbf{P} = Pu \cos \beta. \tag{1.2}$$

This equation represents the definition of the dot product and may be applied to any two vectors. Since the projection of a vector $\mathbf{u} = \mathbf{v} + \mathbf{w}$ on the direction of $\mathbf{P}$ is equal to the sum of the projections of $\mathbf{v}$ and $\mathbf{w}$, it is evident that the dot product has the distributive property:

$$\mathbf{P} \cdot (\mathbf{v} + \mathbf{w}) = \mathbf{P} \cdot \mathbf{v} + \mathbf{P} \cdot \mathbf{w}.$$

When any one of the unit vectors $\mathbf{i}_x$, $\mathbf{i}_y$, $\mathbf{i}_z$ is dot-multiplied with itself, the angle $\beta$ of (1.2) is zero, hence

$$\mathbf{i}_x \cdot \mathbf{i}_x = \mathbf{i}_y \cdot \mathbf{i}_y = \mathbf{i}_z \cdot \mathbf{i}_z = 1.$$

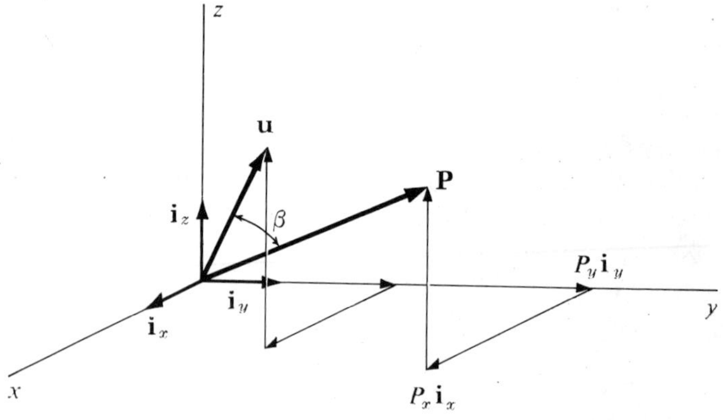


FIGURE 1.1 *Vectors in cartesian coordinates.*

If, on the other hand, two different unit vectors are multiplied with each other, they are at right angles and $\cos\beta = 0$, hence

$$\mathbf{i}_x \cdot \mathbf{i}_y = \mathbf{i}_y \cdot \mathbf{i}_z = \mathbf{i}_z \cdot \mathbf{i}_x = 0.$$

These relations may be combined into a single one:

$$\mathbf{i}_m \cdot \mathbf{i}_n = \delta_{mn}, \qquad m, n = x, y, z \tag{1.3}$$

if we introduce the *Kronecker delta*, $\delta_{mn}$, by the equations

$$\begin{aligned} \delta_{mn} &= 1 \quad \text{if} \quad m = n, \\ \delta_{mn} &= 0 \quad \text{if} \quad m \neq n. \end{aligned} \tag{1.4}$$

We write the dot product of the right-hand sides of (1.1a, b):

$$\mathbf{P} \cdot \mathbf{u} = (P_x\mathbf{i}_x + P_y\mathbf{i}_y + P_z\mathbf{i}_z) \cdot (u_x\mathbf{i}_x + u_y\mathbf{i}_y + u_z\mathbf{i}_z).$$

When we multiply the two sums term by term, we encounter all the possible combinations of $m$ and $n$ in (1.3). Because of (1.4), only three of the nine products survive and we have

$$\mathbf{P} \cdot \mathbf{u} = P_x u_x + P_y u_y + P_z u_z, \tag{1.5}$$

a well-known formula of elementary vector algebra.

We try now to repeat this line of thought in a skew coordinate system. To simplify the demonstration, we restrict ourselves to two dimensions (Figure 1.2). We write the work, i.e. the dot product, first as the work done by $P_x\mathbf{i}_x$ plus that done by $P_y\mathbf{i}_y$:

$$\mathbf{P} \cdot \mathbf{u} = W = P_x(u_x + u_y\cos\alpha) + P_y(u_y + u_x\cos\alpha) \tag{1.6a}$$

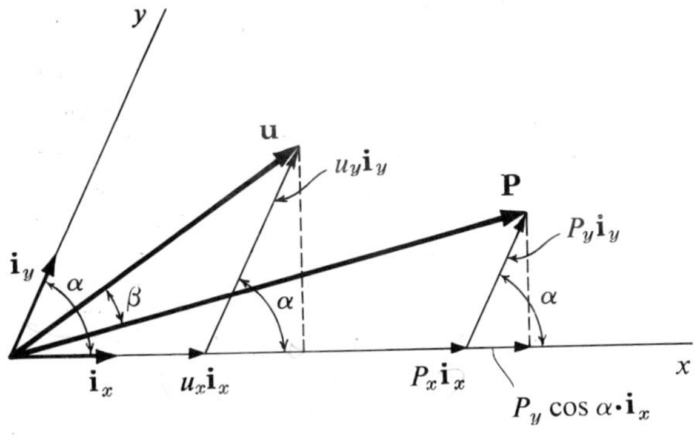


FIGURE 1.2 *Vectors in a skew rectilinear coordinate system.*

and then as the work done by **P** if the displacements $u_x\mathbf{i}_x$ and $u_y\mathbf{i}_y$ occur subsequently:

$$\mathbf{P}\cdot\mathbf{u} = W = u_x(P_x + P_y\cos\alpha) + u_y(P_y + P_x\cos\alpha). \tag{1.6b}$$

Each of these equations can be brought into the form

$$\mathbf{P}\cdot\mathbf{u} = P_x u_x + P_y u_y + (P_x u_y + P_y u_x)\cos\alpha.$$

This has yielded a result, but it has not increased our insight. It is better to leave equations (1.6) as they stand and to realize that we must deal with two different sets of components of each vector: (i) the usual ones, like $P_x$, $P_y$, which are obtained when **P** is made the diagonal of a parallelogram whose sides are parallel to the coordinate axes, and (ii) the components $(P_x + P_y\cos\alpha)$, $(P_y + P_x\cos\alpha)$, which are the normal projections of **P** on the axes $x$ and $y$.

Before we embark upon a closer inspection of these components we introduce the notation which is fundamental for tensor theory and which will be used from now on in this book. Instead of components $P_x$, $P_y$ we write $P^1$, $P^2$, using superscripts, and call these quantities the contravariant components of the vector **P**. For the second set of components we write

$$P_x + P_y\cos\alpha = P_1,$$

$$P_y + P_x\cos\alpha = P_2$$

and call these the covariant components of **P**. The quantities $P^n$ are vector components in the familiar sense of the word. When we multiply them with the unit vectors $\mathbf{i}_x = \mathbf{i}_1$ and $\mathbf{i}_y = \mathbf{i}_2$ and add the products, we obtain the vector **P**:

$$\mathbf{P} = P^1\mathbf{i}_1 + P^2\mathbf{i}_2 = \sum P^n\mathbf{i}_n. \tag{1.7a}$$

The covariant components can be added in a similar manner if we interpret them as shown in Figure 1.3. This figure contains, besides the axes 1 and 2, two other axes, which are at right angles to them. On these we project **P** by the usual parallelogram construction to obtain components of the magnitude

$$\frac{P_x + P_y\cos\alpha}{\sin\alpha} = \frac{P_1}{\sin\alpha}$$

and

$$\frac{P_y + P_x\cos\alpha}{\sin\alpha} = \frac{P_2}{\sin\alpha}.$$

When we interpret them as vectors, they add up to form **P**. We write them

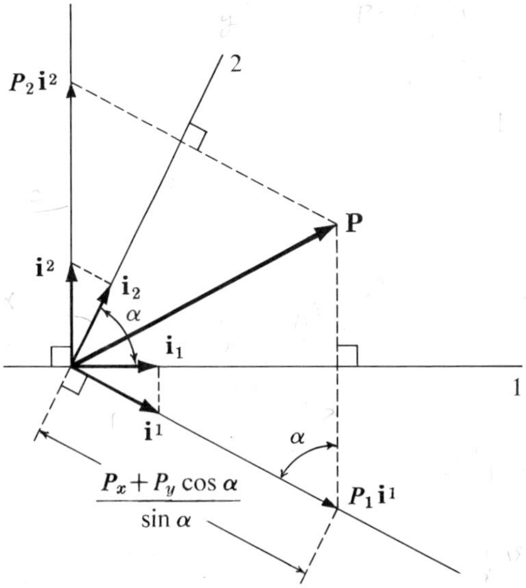


FIGURE 1.3 *Covariant and contravariant components of a vector.*

using reference vectors $\mathbf{i}^1$, $\mathbf{i}^2$, which are not unit vectors, but have the absolute value $1/\sin\alpha$. Then we can write

$$\mathbf{P} = P_1\mathbf{i}^1 + P_2\mathbf{i}^2 = \sum_m P_m\mathbf{i}^m \tag{1.7b}$$

s a second component representation of the vector **P**.

The idea explained here in two dimensions may easily be extended to three (and even more) dimensions. We choose an arbitrary set of three unit ectors $\mathbf{i}_1$, $\mathbf{i}_2$, $\mathbf{i}_3$, called a reference frame. Then we resolve an arbitrary ector $\mathbf{v}$ in the usual way into components along the directions of these unit ectors and write them as $v^n\mathbf{i}_n$, $n = 1, 2, 3$. The vector $\mathbf{v}$ is then the sum of these contravariant components

$$\mathbf{v} = \sum_n v^n\mathbf{i}_n. \tag{1.8}$$

Next, we choose vectors $\mathbf{i}^m$, which satisfy the condition

$$\mathbf{i}^m \cdot \mathbf{i}_n = \delta_n^m, \tag{1.9}$$

where $\delta_n^m$ is another way of writing the Kronecker symbol $\delta_{mn}$ defined in (1.4). Each of the vectors $\mathbf{i}^m$ defined by (1.9) is at right angles to the vectors $\mathbf{i}_n$ with $n \neq m$ and of such magnitude that its absolute value $|\mathbf{i}^n|$ is the reciprocal of cos $(\mathbf{i}_n, \mathbf{i}^n)$. We may now resolve $\mathbf{v}$ in components in the directions of the vectors $\mathbf{i}^m$ and write

$$\mathbf{v} = \sum v_m \mathbf{i}^m. \tag{1.10}$$

Since, in general, $|\mathbf{i}^m| \neq 1$, the covariant components $v_m$ are not simply the absolute values of the component vectors $v_m \mathbf{i}^m$.

When we now consider any two vectors $\mathbf{u}$ and $\mathbf{v}$, we may resolve one of them into contravariant components according to (1.8) and the other one into covariant ones according to (1.10):

$$\mathbf{u} = \sum_n u^n \mathbf{i}_n, \qquad \mathbf{v} = \sum_m v_m \mathbf{i}^m.$$

The dot product is then

$$\mathbf{u} \cdot \mathbf{v} = \sum_n \sum_m u^n v_m \mathbf{i}_n \cdot \mathbf{i}^m = \sum_n \sum_m u^n v_m \delta_n^m. \tag{1.11}$$

The double sum contains all possible combinations of $n$ and $m$, nine terms all together. However, only in the three terms for which $n = m$, does the Kronecker symbol $\delta_n^m = 1$, while for the other six it equals zero. We may, therefore, write

$$\mathbf{u} \cdot \mathbf{v} = \sum_n u^n v_n = u^1 v_1 + u^2 v_2 + u^3 v_3, \tag{1.12}$$

which shows that also in skew rectilinear coordinates the formula for the dot product is as simple as (1.5), if for one vector we use the contravariant components and for the other the covariant ones.

We may now do the final step to build up the notation to be used with vectors and tensors. It will turn out that we always have to deal with sums over some index which appears twice in each term, once as a superscript in a contravariant component and once as a subscript indicating a covariant component. We shall in all these cases omit the summation sign and use the

SUMMATION CONVENTION: Whenever the same Latin letter (say $n$) appears in a product once as a subscript and once as a superscript, it is understood that this means a sum of all terms of this kind (i.e. for $n=1$, 2, 3).

With this convention we rewrite (1.12) as

$$\mathbf{u} \cdot \mathbf{v} = u^n v_n \tag{1.13}$$

and (1.11) as

$$\mathbf{u} \cdot \mathbf{v} = u^n v_m \mathbf{i}_n \cdot \mathbf{i}^m = u^n v_m \delta_n^m, \tag{1.14}$$

implying in this case a summation over all $n$ and over all $m$.

Since in the result of such a summation the summation index no longer appears, it does not matter which letter we use for it. Such an index is called a dummy index, and when necessary, we may change the letter used for it from one equation to the next or from the left-hand side to the right-hand side of the same equation. It will often be necessary to do so because we must avoid making the summation convention unclear by using the same letter for two sums.

## 1.2. Base Vectors, Metric Tensor

In (1.7a) we used unit vectors $\mathbf{i}_n$ as a base for defining the contravariant components, but in (1.7b) we found it necessary to choose vectors $\mathbf{i}^n$, which do not have unit magnitude. We broaden our experience by considering a vector in a polar coordinate system, Figure 1.4. As a specimen vector we choose a line element $d\mathbf{s}$. Defining unit vectors $\mathbf{i}_1, \mathbf{i}_2$ in the direction of increasing coordinates, we can write

$$d\mathbf{s} = \mathbf{i}_1\, dr + \mathbf{i}_2 r\, d\theta. \tag{1.15}$$

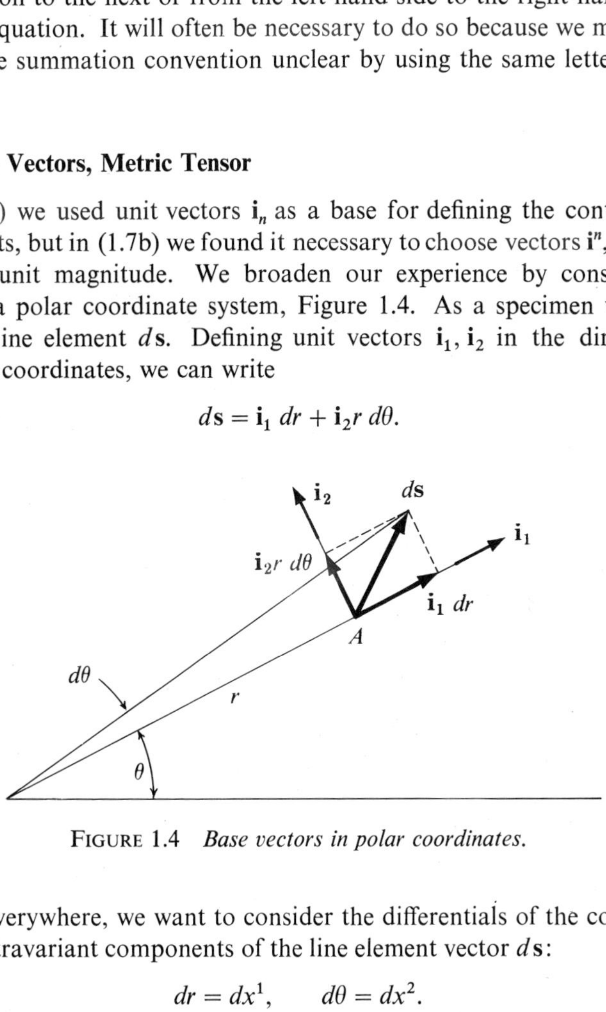


FIGURE 1.4 *Base vectors in polar coordinates.*

Here, as everywhere, we want to consider the differentials of the coordinates as the contravariant components of the line element vector $d\mathbf{s}$:

$$dr = dx^1, \qquad d\theta = dx^2.$$

It is then necessary that, instead of unit vectors, we use the coefficients of these differentials in (1.15) as base vectors:

$$\mathbf{g}_1 = \mathbf{i}_1, \qquad \mathbf{g}_2 = \mathbf{i}_2 r.$$

We call them the contravariant *base vectors* and rewrite (1.15) in the form

$$d\mathbf{s} = \mathbf{g}_1\, dx^1 + \mathbf{g}_2\, dx^2 = \mathbf{g}_i\, dx^i. \tag{1.16}$$

While $\mathbf{g}_1$ is still a unit vector, $\mathbf{g}_2$ has the absolute value $r$ and is not even dimensionless, as the unit vectors are. We see also that, different from rectilinear coordinates, these base vectors are not constant, but depend on the coordinates of the point $A$, for which they have been defined. The directions of $\mathbf{g}_1$ and $\mathbf{g}_2$ depend on $\theta$ and the magnitude of $\mathbf{g}_2$ depends on $r$.

We generalize (1.16) by extending it to an arbitrary (possibly curvilinear) three-dimensional coordinate system $x^i$ $(i = 1, 2, 3)$. At any point $A$ we choose three vectors $\mathbf{g}_i$ of such direction and magnitude that the line element vector

$$d\mathbf{s} = \mathbf{g}_i \, dx^i. \tag{1.17}$$

Now consider the position vector $\mathbf{r}$ leading from a fixed point $O$ (possibly the origin of the coordinates) to the point $A$. The line element $d\mathbf{s}$ is the increment of $\mathbf{r}$ connected with the transition to an adjacent point, $d\mathbf{s} = d\mathbf{r}$. We can write this increment in the form

$$d\mathbf{r} = \frac{\partial \mathbf{r}}{\partial x^i} dx^i,$$

where again the summation convention is to be applied (accepting the superscript in the denominator in lieu of the required subscript). Comparing this expression with (1.17), we see that

$$\mathbf{g}_i = \frac{\partial \mathbf{r}}{\partial x^i}. \tag{1.18}$$

We apply the base vectors $\mathbf{g}_i$ defined by (1.17) or (1.18) to all vectors associated with the point $A$. As an example, a force $\mathbf{P}$ acting at this point is written as

$$\mathbf{P} = \mathbf{g}_i P^i; \tag{1.19}$$

any of the components $P^i$ has the dimension of a force if the corresponding $\mathbf{g}_i$ is dimensionless [as $\mathbf{g}_1$ in (1.16)] and otherwise has such a dimension that its product with $\mathbf{g}_i$ is a force.

A second set of base vectors $\mathbf{g}^j$ is defined by an equation similar to (1.9):

$$\mathbf{g}_i \cdot \mathbf{g}^j = \delta_i^j. \tag{1.20}$$

Each vector $\mathbf{g}^j$ is at right angles to all vectors $\mathbf{g}_i$ for which $i \neq j$ and has such magnitude (and such a dimension) that its dot product with $\mathbf{g}_j$ equals unity. This defines completely the vectors $\mathbf{g}^j$, which are called the contravariant base vectors. They may be used to define covariant components $P_j$ of any vector $\mathbf{P}$:

$$\mathbf{P} = \mathbf{g}^j P_j. \tag{1.21}$$

When we apply the definitions (1.19) and (1.21) to any two vectors **u** and **v**, we may write their dot product as

$$\mathbf{u} \cdot \mathbf{v} = u^i \mathbf{g}_i \cdot v_j \mathbf{g}^j = u^i v_j \mathbf{g}_i \cdot \mathbf{g}^j = u^i v_j \delta_i^j = u^i v_i \qquad (1.22a)$$

since

$$v_j \delta_i^j = v_i$$

and in the alternate form

$$\mathbf{u} \cdot \mathbf{v} = u_i \mathbf{g}^i \cdot v^j \mathbf{g}_j = u_i v^j \delta_j^i = u_i v^i. \qquad (1.22b)$$

Every vector can be resolved into covariant or into contravariant components. When we try to write the covariant base vector $\mathbf{g}_1$ in contravariant components, we have

$$\mathbf{g}_1 = \mathbf{g}_1 \cdot 1 + \mathbf{g}_2 \cdot 0 + \mathbf{g}_3 \cdot 0,$$

i.e. a triviality. No matter what the actual magnitude of $\mathbf{g}_1$ is, it always has the components (1, 0, 0) in the system of base vectors $\mathbf{g}_i$. However, when we resolve a covariant base vector into covariant components, we are led to a set of new, important quantities:

$$\mathbf{g}_i = g_{ij} \mathbf{g}^j. \qquad (1.23a)$$

The entity of the nine quantities $g_{ij}$ thus defined is called the *metric tensor* and the individual $g_{ij}$ are its covariant components. The meaning which stands behind this terminology will become clear when we study the tensor concept (see p. 15).

In analogy to (1.23a), we may resolve $\mathbf{g}^i$ into contravariant components,

$$\mathbf{g}^i = g^{ij} \mathbf{g}_j \qquad (1.23b)$$

and thus define contravariant components of the metric tensor.

Let us now consider dot products of base vectors of the same set:

$$\mathbf{g}_i \cdot \mathbf{g}_j = g_{ik} \mathbf{g}^k \cdot \mathbf{g}_j = g_{ik} \delta_j^k = g_{ij} \qquad (1.24a)$$

or

$$\mathbf{g}^i \cdot \mathbf{g}^j = g^{ik} \mathbf{g}_k \cdot \mathbf{g}^j = g^{ik} \delta_k^j = g^{ij}. \qquad (1.24b)$$

Since the two factors in a dot product may be interchanged, it follows that

$$g_{ij} = g_{ji}, \qquad g^{ij} = g^{ji}. \qquad (1.25)$$

Equations (1.24) may be used as the definitions of $g_{ij}$ and $g^{ij}$. If this is done, (1.23) must be derived from them. This can be done in the following way: Tentatively, let

$$\mathbf{g}_i = a_{ij} \mathbf{g}^j$$

and dot-multiply both sides of this equation by $\mathbf{g}_k$. This yields

$$\mathbf{g}_i \cdot \mathbf{g}_k = a_{ij}\,\mathbf{g}^j \cdot \mathbf{g}_k = a_{ij}\,\delta^j_k = a_{ik}$$

and, because of (1.24a),

$$a_{ik} = g_{ik}.$$

Next, let us consider the product of base vectors from opposite systems. From (1.20) we know that $\mathbf{g}_i \cdot \mathbf{g}^j = \delta^j_i$ and, on the other hand, we have

$$\mathbf{g}_i \cdot \mathbf{g}^j = g_{ik}\,\mathbf{g}^k \cdot g^{jl}\,\mathbf{g}_l = g_{ik}\,g^{jl}\delta^k_l = g_{ik}\,g^{jk}.$$

Comparing both statements, we obtain the result

$$g_{ik}\,g^{jk} = \delta^j_i. \tag{1.26}$$

For a fixed value of $j$ and for $i = 1, 2, 3$ this yields three component equations

$$g_{11}g^{j1} + g_{12}\,g^{j2} + g_{13}\,g^{j3} = \delta^j_1,$$
$$g_{21}g^{j1} + g_{22}\,g^{j2} + g_{23}\,g^{j3} = \delta^j_2,$$
$$g_{31}g^{j1} + g_{32}\,g^{j2} + g_{33}\,g^{j3} = \delta^j_3,$$

which may be solved for the $g^{jk}$, $k = 1, 2, 3$ if the $g_{ik}$ are known. Conversely, (1.26) may also be used to calculate the $g_{ik}$ from known $g^{jk}$.

As a last step in exploring the properties of the $g_{ij}$, we consider a line element vector $d\mathbf{s}$ as in (1.17) and dot-multiply it by itself:

$$d\mathbf{s} \cdot d\mathbf{s} = \mathbf{g}_i\,dx^i \cdot \mathbf{g}_j\,dx^j = g_{ij}\,dx^i\,dx^j. \tag{1.27}$$

This equation is often used to derive expressions for the $g_{ij}$ in terms of the coordinates. Its left-hand side is the square of the line element, and it is usually easy to express it as a quadratic form in the coordinate differentials $dx^i$. Its coefficients are the $g_{ij}$. In this quadratic form it is, of course, not possible to distinguish between the coefficients of $dx^1\,dx^2$ and $dx^2\,dx^1$, i.e. between $g_{12}$ and $g_{21}$; but because of (1.25) this is not necessary since $g_{12}$ and $g_{21}$ are each half of the total coefficient of the product of these two differentials. Once the $g_{ij}$ have been found, (1.26) can be used to calculate the $g^{ij}$.

The $g_{ij}$ and $g^{ij}$ may be used to express the covariant and the contravariant components of a vector in terms of each other. Take an arbitrary vector $\mathbf{u}$ and write

$$\mathbf{u} = u^i\mathbf{g}_i.$$

Using (1.23a), we may write instead

$$\mathbf{u} = u^i g_{ij}\,\mathbf{g}^j,$$

and this must be the same as $u_j \mathbf{g}^j$. However, when we write the equation

$$u^i g_{ij} \mathbf{g}^j = u_j \mathbf{g}^j,$$

we cannot simply "cancel the factor" $\mathbf{g}^j$ on both sides, since these expressions are not simple products but rather sums of three products containing the factors $\mathbf{g}^1$, $\mathbf{g}^2$, $\mathbf{g}^3$, respectively. To get rid of the factor $\mathbf{g}^j$, we must use the following procedure. We dot-multiply both sides by $\mathbf{g}_k$. This yields on the left-hand side

$$u^i g_{ij} \mathbf{g}^j \cdot \mathbf{g}_k = u^i g_{ij} \delta^j_k = u^i g_{ik}$$

and on the right-hand side

$$u_j \mathbf{g}^j \cdot \mathbf{g}_k = u_j \delta^j_k = u_k,$$

and now we can equate the results:

$$u^i g_{ik} = u_k. \tag{1.28a}$$

In the same way one shows that

$$u_i g^{ik} = u^k. \tag{1.28b}$$

The operation described by (1.28) is known as lowering or raising an index.

With the help of these equations we may obtain two more forms of the dot product:

$$\mathbf{u} \cdot \mathbf{v} = u_i v^i = u_i v_j g^{ij} = u^j v^i g_{ij}. \tag{1.29}$$

Here is a good place to pause and look back at what we have done. To define the components of a vector, we need base vectors $\mathbf{g}_i$, $\mathbf{g}^j$, and since these base vectors depend on the coordinates, for each vector we must choose a definite point from which the base vectors are to be taken. Such a choice cannot be meaningful unless the physical quantity, which the vector represents, is attached to a definite point in space. For a force, this is the point where it is applied to a body. For a velocity in a flow field, it is the location of a specific particle at a specific time. For a line element $d\mathbf{s}$, it is the point where this infinitesimal quantity is supposed to converge to zero. However, a position vector $\mathbf{r}$, as we used it on page 8, is not associated with one point, but with two at a finite distance. Therefore, we cannot resolve it into covariant or contravariant components unless we arbitrarily specify from which point we will take them. One of the essential difficulties of the theory of large elastic deformations is that the displacement vector is not infinitesimal and has to be considered with respect to two sets of base vectors—one located at one end in the undeformed body and the other at the other end in the deformed body.

Let us also have another look at the technique of handling the summation convention. In the second member of (1.27) we have two independent sums separated by the multiplication dot. It is necessary to use two different letters, $i$ and $j$, for the dummy indices to make clear what belongs together in one sum or the other. We may then shift the position of the factors and combine them as suitable, and this has been done in the third member of this equation.

In many equations we find indices which are not dummies. They occur only once in each term and are either all subscripts or all superscripts, as for example, the subscript $k$ in (1.28a) or the superscripts $i$, $j$ in (1.24b). This means that such an equation is valid, separately, for any value this index may assume. Therefore, (1.28a) represents three equations, written for $k = 1, 2, 3$, and (1.24b) represents nine equations for the nine possible combinations of $i$ and $j$.

## 1.3. Coordinate Transformation

Let us consider two reference frames, the "old" frame $\mathbf{g}_i$ and the "new" frame $\mathbf{g}_{i'}$, where $i' = 1', 2', 3'$. We assume that a vector is known by its components $v^i$ in the old reference frame and we want to calculate its components in the new one.

In each of the two frames we can define a set of contravariant base vectors $\mathbf{g}^j$, $\mathbf{g}^{j'}$, which satisfy (1.20):

$$\mathbf{g}_i \cdot \mathbf{g}^j = \delta_i^j, \qquad \mathbf{g}_{i'} \cdot \mathbf{g}^{j'} = \delta_{i'}^{j'}. \tag{1.30a, b}$$

The new base vectors may be resolved in components with respect to the old base vectors of the same variance:

$$\mathbf{g}_{i'} = \beta_{i'}^j \mathbf{g}_j, \qquad \mathbf{g}^{i'} = \beta_j^{i'} \mathbf{g}^j. \tag{1.31a, b}$$

These equations define two sets of nine quantities each, which express a relation between the two reference frames. We introduce (1.31) with suitable changes of the dummy indices into (1.30b):

$$\delta_{i'}^{k'} = \mathbf{g}_{i'} \cdot \mathbf{g}^{k'} = \beta_{i'}^j \mathbf{g}_j \cdot \beta_l^{k'} \mathbf{g}^l = \beta_{i'}^j \beta_l^{k'} \mathbf{g}_j \cdot \mathbf{g}^l = \beta_{i'}^j \beta_l^{k'} \delta_j^l$$

and after summation over $l$:

$$\beta_{i'}^j \beta_j^{k'} = \delta_{i'}^{k'}. \tag{1.32a}$$

Since this equation can be written separately for any $i'$ and $k'$, it represents nine equations between the various $\beta$. We may write them in matrix form:

$$\begin{bmatrix} \beta_{1'}^1 & \beta_{1'}^2 & \beta_{1'}^3 \\ \beta_{2'}^1 & \beta_{2'}^2 & \beta_{2'}^3 \\ \beta_{3'}^1 & \beta_{3'}^2 & \beta_{3'}^3 \end{bmatrix} \begin{bmatrix} \beta_1^{1'} & \beta_1^{2'} & \beta_1^{3'} \\ \beta_2^{1'} & \beta_2^{2'} & \beta_2^{3'} \\ \beta_3^{1'} & \beta_3^{2'} & \beta_3^{3'} \end{bmatrix} = \begin{bmatrix} 1 & 0 & 0 \\ 0 & 1 & 0 \\ 0 & 0 & 1 \end{bmatrix}.$$

In matrix language this means that the matrices formed of the $\beta^j_{i'}$ and the $\beta^{k'}_j$ are reciprocals of each other. If one of them is known, the elements of the other one can be calculated. One easily derives a counterpart to (1.32a):

$$\beta^{j'}_i \beta^k_{j'} = \delta^k_i . \tag{1.32b}$$

Now let us multiply both sides of (1.31a) by $\beta^{i'}_k$. This involves, of course, application of the summation convention to $i'$ and because of (1.32b) yields

$$\beta^{i'}_k \mathbf{g}_{i'} = \beta^{i'}_k \beta^j_{i'} \mathbf{g}_j = \delta^j_k \mathbf{g}_j = \mathbf{g}_k , \tag{1.33a}$$

i.e. an expression for the old base vectors in terms of the new ones. Similarly, one can show that

$$\mathbf{g}^k = \beta^k_{i'} \mathbf{g}^{i'} . \tag{1.33b}$$

We may now define components of a vector $\mathbf{v}$ in both reference frames:

$$\mathbf{v} = v_i \mathbf{g}^i = v^i \mathbf{g}_i = v_{i'} \mathbf{g}^{i'} = v^{i'} \mathbf{g}_{i'} . \tag{1.34}$$

To find a relationship between $v_i$ and $v_{i'}$, we make use of (1.33b):

$$\mathbf{v} = v_i \mathbf{g}^i = v_i \beta^i_{k'} \mathbf{g}^{k'} = v_{k'} \mathbf{g}^{k'} .$$

Again, we cannot simply "cancel the factor" $\mathbf{g}^{k'}$ from the third and fourth members, but we may dot-multiply both sides by $\mathbf{g}_{j'}$:

$$v_i \beta^i_{k'} \mathbf{g}^{k'} \cdot \mathbf{g}_{j'} = v_{k'} \mathbf{g}^{k'} \cdot \mathbf{g}_{j'} ,$$

$$\therefore \quad v_i \beta^i_{k'} \delta^{k'}_{j'} = v_{k'} \delta^{k'}_{j'} ,$$

$$\therefore \quad v_i \beta^i_{j'} = v_{j'} . \tag{1.35a}$$

Multiplying both sides by $\beta^{j'}_k$, we obtain on the left

$$v_i \beta^i_{j'} \beta^{j'}_k = v_i \delta^i_k = v_k$$

and hence

$$v_k = v_{j'} \beta^{j'}_k . \tag{1.35b}$$

Equations (1.35) express the new covariant components in terms of the old ones and vice versa. Similarly, one derives a pair of formulas for the contravariant components:

$$v^{j'} = v^i \beta^{j'}_i , \qquad v^j = v^{i'} \beta^j_{i'} . \tag{1.35c, d}$$

We may summarize (1.33) and (1.35) in the following form: Any quantity which has a *subscript*, like $\mathbf{g}_i$, $v_i$, must be multiplied by a $\beta^i_{j'}$ to transform it to the new reference frame, and whatever has a *superscript*, like $\mathbf{g}^i$, $v^i$, needs a $\beta^{j'}_i$. We arbitrarily take the base vectors $\mathbf{g}_i$ as the standard of comparison, and all quantities which vary in the same way are called *covariant* and those which vary in the opposite way are called *contravariant*.

In a coordinate system $x^i$ we may write the line element $d\mathbf{r}$ in the following two forms:

$$d\mathbf{r} = \mathbf{g}_i \, dx^i = \beta_i^{k'} \mathbf{g}_{k'} \, dx^i$$

and

$$d\mathbf{r} = \mathbf{g}_{k'} \, dx^{k'} = \mathbf{g}_{k'} \frac{\partial x^{k'}}{\partial x^i} \, dx^i.$$

Equating the right-hand sides again yields an equation in which we would like to cancel a factor:

$$\beta_i^{k'} \mathbf{g}_{k'} \, dx^i = \mathbf{g}_{k'} \frac{\partial x^{k'}}{\partial x^i} \, dx^i.$$

To legalize the dropping of $dx^i$, we proceed in the following way: The $dx^i$ ($i = 1, 2, 3$) are the components of $d\mathbf{r}$, and the equation is valid for any choice of $d\mathbf{r}$. Now choose a line element which points in the direction 1. Then $dx^2 = dx^3 = 0$, and since the sums over $i$ have only one term $i = 1$, the factor $dx^1$ can be cancelled. Then consider line elements in directions 2 and 3 to find that, for every $i$ separately, the relation

$$\beta_i^{k'} \mathbf{g}_{k'} = \mathbf{g}_{k'} \frac{\partial x^{k'}}{\partial x^i}$$

holds true. To get rid of the factor $\mathbf{g}_{k'}$, we proceed as we have done already on similar occasions. We dot-multiply by $\mathbf{g}^{j'}$, find Kronecker deltas and perform summations to find that

$$\beta_i^{j'} = \frac{\partial x^{j'}}{\partial x^i}. \tag{1.36a}$$

Similarly,

$$\beta_{j'}^{i} = \frac{\partial x^{i}}{\partial x^{j'}}. \tag{1.36b}$$

Equations (1.36) can be used to calculate the transformation coefficients $\beta_i^{j'}$ and $\beta_{j'}^{i}$ by differentiating the equations which express one set of coordinates in terms of the other. Often it is difficult to invert the relations $x^i = x^i(x^{j'})$ or $x^{j'} = x^{j'}(x^i)$ and only one of (1.36) can be applied. Then (1.32) may be used to find the opposite set of $\beta$ coefficients.

We obtain an interesting result when we ask how the $g_{ij}$ transform into the new coordinates. The answer may be found in different ways. One is is to write (1.24) in the new coordinates and then to apply (1.31):

$$g_{i'j'} = \mathbf{g}_{i'} \cdot \mathbf{g}_{j'} = \beta_{i'}^{k} \mathbf{g}_k \cdot \beta_{j'}^{l} \mathbf{g}_l = \beta_{i'}^{k} \beta_{j'}^{l} g_{kl}, \tag{1.37a}$$

$$g^{i'j'} = \mathbf{g}^{i'} \cdot \mathbf{g}^{j'} = \beta_k^{i'} \mathbf{g}^k \cdot \beta_l^{j'} \mathbf{g}^l = \beta_k^{i'} \beta_l^{j'} g^{kl}. \tag{1.37b}$$

The $g_{ij}$ transform similarly to the covariant vector components in (1.35a), but we need two $\beta$ factors, one for each subscript, and the $g^{ij}$ show what we may call double contravariant behavior. We may now understand why the $g_{ij}$ and $g^{ij}$ are called components of the metric tensor. Metric because of (1.27); they correlate the actual length of the line element $(d\mathbf{s} \cdot d\mathbf{s})^{1/2}$ and the increments $dx^i$ of the coordinates and thus establish the metric of the space of these coordinates. Tensor because of (1.37), since we shall see in the next section that this law of transformation is the essential property of a tensor. Covariant and contravariant because of the type of $\beta$ occurring in each of (1.37).

There is a second way of deriving (1.37), and since it contains an important technique of reasoning we shall demonstrate it here. We start from (1.27):

$$d\mathbf{s} \cdot d\mathbf{s} = g_{i'j'}\, dx^{i'}\, dx^{j'} = g_{ij}\, dx^i\, dx^j.$$

Since $dx^i$ are vector components, (1.35d) may be applied to them:

$$g_{i'j'}\, dx^{i'}\, dx^{j'} = g_{ij}\, \beta^i_{i'}\, dx^{i'} \beta^j_{j'}\, dx^{j'}.$$

Since this equation holds for the components of any line element vector $d\mathbf{s}$, we first choose one for which only $dx^{1'} \neq 0$ and thus prove (1.37) for $i' = j' = 1'$. Then we do the same for $i' = j' = 2'$ and $= 3'$. Then we use a line element for which only $dx^{3'} = 0$ and after cancelling from both sides of the equation what is already known to be equal, we are left with

$$g_{1'2'}\, dx^{1'}\, dx^{2'} + g_{2'1'}\, dx^{2'}\, dx^{1'} = g_{ij}(\beta^i_{1'}\, \beta^j_{2'}\, dx^{1'}\, dx^{2'} + \beta^i_{2'}\, \beta^j_{1'}\, dx^{2'}\, dx^{1'}).$$

We may cancel the differentials and in the last term we may interchange $i$ and $j$:

$$g_{1'2'} + g_{2'1'} = g_{ij}\, \beta^i_{1'}\, \beta^j_{2'} + g_{ji}\, \beta^i_{1'}\, \beta^j_{2'},$$

whence, because of $g_{1'2'} = g_{2'1'}$ and $g_{ij} = g_{ji}$, (1.37) is seen to be true for $g_{1'2'}$. This can be continued to cover the remaining components $g_{2'3'}$ and $g_{1'3'}$.

We now know two techniques for "cancelling a factor" in an equation: (1) We make use of the general validity of the equation and go through a sequence of special cases, or (2) when the factor to be cancelled is a base vector, we dot-multiply by another base vector to produce a Kronecker delta.

## 1.4. Tensors

We shall now generalize the result we obtained when transforming the components of the metric tensor. Consider two vectors $\mathbf{a}$ and $\mathbf{b}$, which, in a reference frame $\mathbf{g}^i$, have the components $a_i$ and $b_i$. In a second (the "new") reference frame $g^{i'}$ they have components $a_{i'}$ and $b_{i'}$. We form the products

$$c_{ij} = a_i b_j \qquad \text{and} \qquad c_{i'j'} = a_{i'} b_{j'}. \tag{1.38a, b}$$

Such products may represent physical quantities, but at present our thinking is merely formal and not attached to any physical visualization. We ask how the nine quantities $c_{i'j'}$ depend on the nine $c_{ij}$.

The answer comes, of course, from the transformation formula (1.35a). We use it to express $a_{i'}$ and $b_{j'}$ in terms of $a_i$ and $b_j$ and find

$$c_{i'j'} = a_i \beta^i_{i'} b_j \beta^j_{j'} = c_{ij} \beta^i_{i'} \beta^j_{j'}, \tag{1.39a}$$

i.e. the same relation as (1.37a). Similarly we can use (1.35b) to find that

$$c_{ij} = a_{i'} \beta^{i'}_i b_{j'} \beta^{j'}_j = c_{i'j'} \beta^{i'}_i \beta^{j'}_j. \tag{1.39b}$$

We may also use the contravariant components of **a** and **b** and define

$$c^{ij} = a^i b^j \qquad \text{and} \qquad c^{i'j'} = a^{i'} b^{j'}, \tag{1.40}$$

and we can derive the relations

$$c^{i'j'} = c^{ij} \beta^{i'}_i \beta^{j'}_j, \qquad c^{ij} = c^{i'j'} \beta^i_{i'} \beta^j_{j'}. \tag{1.41a,b}$$

In the terminology introduced in connection with (1.37), $c_{ij}$ shows double covariant behavior and $c^{ij}$ is double-contravariant. We call them the covariant and the contravariant components of a *tensor*. The physical entity which these components represent is not as easily visualized as a vector. Since there are nine components $c_{ij}$, an arrow has not enough free parameters and in this book we shall not make any attempt at the visualization of tensors. We shall, however, get familiar with a number of physical quantities which are tensors.

We may also define tensor components of mixed variance, e.g.

$$c^{i}_{\cdot j} = a^i b_j \qquad \text{or} \qquad c_i^{\cdot j} = a_i b^j. \tag{1.42}$$

It may be left to the reader to show that

$$c_{i'}^{\cdot j'} = c_i^{\cdot j} \beta^i_{i'} \beta^{j'}_j, \tag{1.43}$$

i.e. that one $\beta$ of each set is needed to transform $c_i^{\cdot j}$ into the new coordinates. Since, in general,

$$a^i b_j = c^i_{\cdot j} \neq c_j^{\cdot i} = a_j b^i,$$

it is important to indicate which index comes first. The little dot used in these symbols serves to mark a vacant space and is in handwriting more useful than in print.

On the left-hand side of (1.38a) there are nine components $c_{ij}$, but they depend on the six components of the vectors **a** and **b**. It follows that not all the $c_{ij}$ are independent of each other. We arrive at a perfectly general tensor with nine independent components when we replace the right-hand side of (1.38a) by a sum of at least two terms:

$$c_{ij} = a_i b_j + d_i e_j.$$

Since all our conclusions apply to each term on the right-hand side of this equation, all our results are equally applicable to these tensor components $c_{ij}$.

There is a second way to define a tensor $c_{ij}$. We start from a set of nine such quantities. Then the equation

$$a^i c_{ij} = b_j \tag{1.44a}$$

associates with every vector **a** another vector **b**; it is a mapping of all the vectors **a** on all the vectors **b**. When we now change from one pair of reference frames $\mathbf{g}_i, \mathbf{g}^i$ to another pair $\mathbf{g}_{i'}, \mathbf{g}^{i'}$, we want to find quantities $c_{i'j'}$ such that the equation

$$a^{i'} c_{i'j'} = b_{j'} \tag{1.44b}$$

connects the same pair of vectors **a** and **b**. We apply the transformations (1.35b, d) to (1.44a):

$$a^{i'} \beta^i_{i'} c_{ij} = b_{k'} \beta^{k'}_j.$$

When we multiply both sides of this equation by $\beta^j_{j'}$, we find on the right-hand side

$$b_{k'} \beta^{k'}_j \beta^j_{j'} = b_{k'} \delta^{k'}_{j'} = b_{j'}$$

and, hence,

$$a^{i'} \beta^i_{i'} \beta^j_{j'} c_{ij} = b_{j'};$$

we see that (1.44b) is satisfied for all pairs $a^{i'}$, $b_{j'}$ if and only if (1.39a) holds.

We now consider quantities $c_i^{\cdot k}$ found from the products of the components of two tensors:

$$c_i^{\cdot k} = a_{ij} b^{jk}, \qquad c_{i'}^{\cdot k'} = a_{i'j'} b^{j'k'}. \tag{1.45a, b}$$

We assume that $a_{ij}$ and $a_{i'j'}$ are the components of the same tensor in two reference terms, i.e. that they are connected by the transformation (1.39) and that similarly (1.41) apply for $b^{jk}$ and $b^{j'k'}$. We ask whether also $c_i^{\cdot k}$ are tensor components, i.e. whether they and $c_{i'}^{\cdot k'}$ are related through (1.43). We find the answer by introducing (1.39a) and (1.41a) on the right-hand side of (1.45b):

$$\begin{aligned} c_{i'}^{\cdot k'} &= a_{ij} \beta^i_{i'} \beta^j_{j'} b^{lk} \beta^{j'}_l \beta^{k'}_k = a_{ij} b^{lk} \delta^j_l \beta^i_{i'} \beta^{k'}_k \\ &= a_{ij} b^{jk} \beta^i_{i'} \beta^{k'}_k = c_i^{\cdot k} \beta^i_{i'} \beta^{k'}_k. \end{aligned}$$

This indeed is (1.43).

In (1.38), (1.40) and (1.42) we have defined tensor components of different variance, starting from the same vectors **a** and **b**. We ask how these components are related to each other. Let us start from (1.38a) and apply (1.28a) to one or both of $a_i$ and $b_j$:

$$c_{ij} = a_i b^k g_{kj} = a^l g_{li} b^k g_{kj}.$$

Comparison with (1.40) and (1.42) shows that this can be written

$$c_{ij} = c_i^{\cdot k} g_{kj} = c^{lk} g_{li} g_{kj} \tag{1.46}$$

and that the procedure of raising or lowering an index, which we derived for vector components on page 11, applies also to tensor components. The comparison also shows that we need one factor $g_{ij}$ for each index to be lowered and one factor $g^{ij}$ for each index to be raised. This gives us the right to consider $c_{ij}$, $c_i^{\cdot j}$, $c_{\cdot j}^{i}$, and $c^{ij}$ as different sets of components of one object, the tensor, although this tensor itself (different from the vectors **a**, **b**) will never appear in our equations.

In general, $c_{ij} \neq c_{ji}$, but we shall later meet with many tensors for which the relation

$$c_{ij} = c_{ji} \tag{1.47}$$

holds. Such tensors are called symmetric. Raising one index, we find from (1.47)

$$c_i^{\cdot k} = c_{ij} g^{jk} = c_{ji} g^{jk} = c^k{}_i$$

and we may simply write $c_i^k$ for either $c_i^{\cdot k}$ or $c^k{}_i$. However, even for a symmetric tensor, there is not $c_i^k = c_k^i$. This relation is not only wrong, but it does not even have correct tensor form, having a subscript $i$ on one side and a superscript $i$ on the other.

When we raise both indices in (1.47), we arrive at the relation

$$c^{ij} = c^{ji}, \tag{1.48}$$

which holds for every symmetric tensor.

Occasionally we shall also meet tensors for which

$$c_{ij} = -c_{ji}. \tag{1.49}$$

This implies that $c_{ij} = 0$ when $i = j$. Tensors of this kind are called antimetric or skew-symmetric. A general second-order tensor $c_{ij}$ can always be written as the sum of a symmetric and an antimetric tensor:

$$c_{ij} = a_{ij} + b_{ij} \tag{1.50}$$

with

$$a_{ij} = \tfrac{1}{2}(c_{ij} + c_{ji}) = a_{ji}, \tag{1.51a}$$

$$b_{ij} = \tfrac{1}{2}(c_{ij} - c_{ji}) = -b_{ji} = -\tfrac{1}{2}(c_{ji} - c_{ij}). \tag{1.51b}$$

If $a_{ij}$ is symmetric and $b^{ij}$ is antimetric, the sum

$$a_{ij}b^{ij} = 0 \tag{1.52}$$

as may easily be verified.

Equation (1.26) may be interpreted as raising the second index of $g_{ik}$. The result, which, because of the symmetry of the metric tensor, should be written as $g_i^j$, turns out to be a Kronecker delta:

$$g_i^j = \delta_i^j. \tag{1.53}$$

It represents the mixed-variance components of the metric tensor in all reference frames.

Thus far, we have distinguished between "the vector **a**" and "the vector components $a_i$ or $a^i$." Since the components define the vector completely, we shall, from now on, ease the language and speak of "the vector $a_i$" and in a similar sense of "the tensor $c_{ij}$," which, of course, is identical with "the tensor $c^{ij}$."

The tensors which were introduced by (1.38), (1.44), and (1.45) always had two indices, which we could choose as subscripts or superscripts. We shall call these tensors, when necessary, *tensors of the second order*. We now introduce tensors of higher order by relations like the following:

$$d_{ijk} = a_i b_j c_k \qquad \text{or} \qquad d_{ijk} = a_{ij} b_k \qquad \text{or} \qquad a^i d_{ijk} = b_{jk}. \tag{1.54}$$

Since we may use covariant or contravariant components in the definitions, this tensor of the third order may have a great many different components, e.g. $d_{ijk}$, $d_{i\cdot k}^{\cdot j}$, $d_{\cdot\cdot k}^{ij}$, $d_i^{\cdot jk}$, etc. We leave it to the reader to prove that: (i) all these components are related to each other by equations like (1.46); (ii) these components are transformed to another reference frame by equations similar to (1.39) and (1.41). For raising or lowering every index one $g$-factor is needed, and the transformation needs three $\beta$-factors—the choice between $\beta_{n'}^m$ and $\beta_m^{n'}$ depends upon the variance of the component to be transformed.

All this applies to tensors of any higher order, which may be formed in the same way.

Tensors of higher order may be symmetric or antimetric with respect to a certain pair of subscripts or superscripts. For example, the permutation tensor $\epsilon_{ijk}$, to be defined on page 33, is antimetric with respect to any pair of subscripts:

$$\epsilon_{ijk} = -\epsilon_{jik} = -\epsilon_{ikj} = -\epsilon_{kji}$$

and the elastic modulus $E^{ijlm}$ introduced in (4.11) is symmetric with respect to the pair $ij$ and with respect to the pair $lm$:

$$E^{ijlm} = E^{jilm} = E^{ijml},$$

but a relation $E^{ijlm} = E^{iljm}$ does not hold.

In any tensor we may choose to make one subscript and one superscript equal and to invoke the summation convention. When this is done to a second-order tensor, no free index is left and the result is a simple number, a *scalar*:

$$c_i^{\cdot i} = a.$$

In a third-order tensor $c_i^{\cdot jk}$ we have

$$c_i^{\cdot ik} = a^k, \qquad c_i^{\cdot ji} = b^j,$$

and these are two different vectors. This operation is called *contraction*, and since it always absorbs two indices, it lowers the order of a tensor by two. Vectors are, in this sense, tensors of the first order, and scalars are tensors of zero order.

The contraction may, of course, be applied to products of components, which have tensor character. For example, the second member of (1.46) is one of the possible contractions of the fourth-order tensor $c_i^{\ k} g_{lj}$.

This equation demonstrates another important fact. All its members are tensors of the second order with the free subscripts $i$, $j$, but they contain different numbers of dummy indices, namely, none, one pair $k$, and two pairs $l$, $k$. To make sense, every tensor equation must be an equation between tensors of the same order and of the same variance and with the same symbols for the free indices. It may, for example, have the form

$$a_{ij} b^{jk} = r_{il} s^{lmn} t_n u_m^{\ k}$$

with a free subscript $i$ and a free superscript $k$, thus representing $3 \cdot 3 = 9$ component equations. When everything is assembled on the left-hand side, the equation has the form

$$A_i^{\cdot k} = 0,$$

to be true for $i, k = 1, 2, 3$. When we now transform all quantities to another reference frame, we see that

$$A_{i'}^{\cdot k'} = A_i^{\cdot k} \beta_{i'}^i \beta_k^{k'}$$

is a sum of terms which are all zero, whence

$$A_{i'}^{\cdot k'} = 0.$$

This proves that any tensor equation which holds true in one reference frame is also true in any other frame. This is a very efficient device for deriving physical equations. All that is needed is to derive an equation in cartesian coordinates and to choose the notations so that every quantity is written as a tensor component. We shall see later that this is not always quite as simple as it may look here, but still simple enough to make this device an important tool of tensor analysis.

*Problems*

1.1. In the spherical coordinate system shown in Figure 1.5, let $r = x^1$, $\theta = x^2$, $\phi = x^3$. (a) Starting from an expression for the square of the line element, calculate $g_{ij}$ and $g^{ij}$. (b) What are the covariant components $v_i$ of a vector whose contravariant components $v^j$ are known? (c) Calculate the transformation coefficients $\beta^i_{i'}$ and $\beta^{i'}_i$ which connect these coordinates with cartesian coordinates $x^{i'}$.

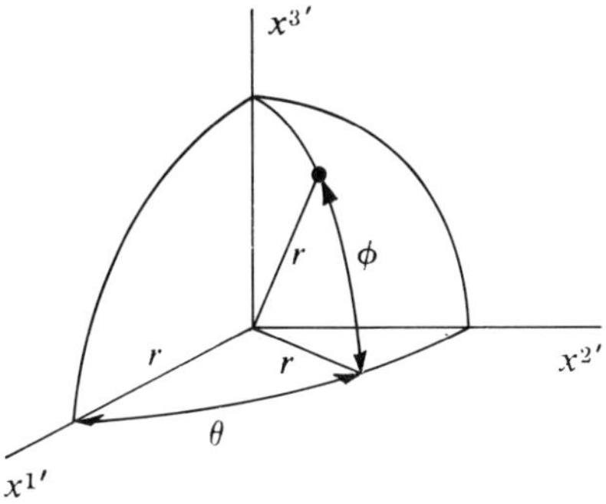


FIGURE 1.5

1.2. In (1.16) the base vectors $\mathbf{g}_i$ have been chosen so that the contravariant components of the line element vector $d\mathbf{s}$ are the increments of the polar coordinates $r$ and $\theta$. Show that the covariant components $dx_i$ of $d\mathbf{s}$ cannot be interpreted as increments of any coordinates $x_i$.

1.3. Replace (1.16) by the equation $d\mathbf{s} = \mathbf{g}^i\, dx_i$ and choose the contravariant base vectors so that $dx_1 = dr$, $dx_2 = d\theta$. What is the relation between these $dx_i$ and those to be derived from the $dx^i$ of the original equation?

## References

There exists a great variety of presentations of tensor analysis. At one end of the gamut are the books serving abstract mathematical purposes, not of direct usefulness in applied mechanics. At the other end one finds chapters on cartesian tensors in many recent books on the theory of elasticity, continuum mechanics, and dynamics. Most of these do not go beyond describing a shorthand notation for writing equations. Here we mention a few books which present general tensors in a form that lends itself to use in applied sciences.

The book by Block [2] is a brief, very readable introduction to the subject. Lass [17] gives a vector analysis in Gibbs formulation in Chapter 2 and an introduction to tensors in Chapter 3. The books by Wills [35], Synge–Schild [31], Coburn [4], and Duschek–Hochrainer [5] are substantial treatises, which may be consulted for further detail beyond the scope of the present book. Hawkins [14] aims directly at the applications. Brillouin [3] starts from very general concepts and gradually narrows them down to the tensor concept used here and elsewhere. In the first chapter, Green–Zerna [13] give a presentation directly aimed at the theory of large elastic deformations.

Some of the authors use Gibbs notation to about the same degree as this book, while others avoid it entirely.

CHAPTER 2

# The Strain Tensor

WE CONSIDER A BODY before and after deformation. In the undeformed body we establish a coordinate system $x^i$ and permanently affix the values of its coordinates to each material point (particle). This means that such points will also be known by the same values $x^i$ even after deformation, although moved to a different place. In other words, the coordinate system undergoes the same deformation as the body. Coordinates used in this way are called particle coordinates or convected coordinates.

Before the deformation we have base vectors $\mathbf{g}_i$ and a metric tensor $g_{ij}$ such that a line element is

$$d\mathbf{s} = \mathbf{g}_i \, dx^i \tag{2.1}$$

and its square

$$d\mathbf{s} \cdot d\mathbf{s} = g_{ij} \, dx^i \, dx^j. \tag{2.2}$$

After deformation, the line element $d\hat{\mathbf{s}}$ connecting the same material points is different in length and direction and may be written

$$d\hat{\mathbf{s}} = \hat{\mathbf{g}}_i \, dx^i. \tag{2.3}$$

The square of the line element is now

$$d\hat{\mathbf{s}} \cdot d\hat{\mathbf{s}} = \hat{g}_{ij} \, dx^i \, dx^j. \tag{2.4}$$

Here $\hat{\mathbf{g}}_i$ is the vector into which the original base vector $\mathbf{g}_i$ has been deformed, and $\hat{g}_{ij}$ is the metric tensor in the deformed coordinate system.

The degree of deformation can be described by the change of the metric tensor, i.e. by the quantities

$$\gamma_{ij} = \hat{g}_{ij} - g_{ij}. \tag{2.5}$$

Both $\hat{g}_{ij}$ and $g_{ij}$ are components of tensors, but in different reference frames, and it is not evident whether their difference $\gamma_{ij}$ is a tensor in the undeformed reference frame $\mathbf{g}_i$. To clarify the situation, we transform everything from the coordinate system $x^i$ to another one, $x^{i'}$, in which

$$d\mathbf{s} = \mathbf{g}_{i'}\, dx^{i'}, \qquad d\hat{\mathbf{s}} = \hat{\mathbf{g}}_{i'}\, dx^{i'}. \tag{2.6}$$

The undeformed base vectors are related by (1.31a), which we rewrite as

$$\mathbf{g}_{i'} = \beta^{i}_{i'}\, \mathbf{g}_i\,,$$

and a similar relation holds in the deformed state:

$$\hat{\mathbf{g}}_{i'} = \beta^{i}_{i'} \hat{\mathbf{g}}_i\,.$$

Since we are using convective coordinates, there is no difference between $x^i$ and $\hat{x}^i$ and from (1.36b) we have

$$\hat{\beta}^{i}_{i'} = \frac{\partial x^i}{\partial x^{i'}} = \beta^{i}_{i'}\,.$$

Therefore, both terms on the right-hand side of (2.5) are transformed in the same way and

$$\gamma_{i'j'} = \gamma_{ij}\, \beta^{i}_{i'}\, \beta^{j}_{j'}\,.$$

This proves the tensor character of $\gamma_{ij}$, and we call the entity of the nine components $\gamma_{ij}$ the *strain tensor*. Later (p. 27) we shall see how these $\gamma_{ij}$ are related to the quantities commonly called strains. Since $g_{ij}$ and $\hat{g}_{ij}$ are symmetric, the same holds for the strain tensor:

$$\gamma_{ij} = \gamma_{ji}\,. \tag{2.7}$$

Equation (2.5) derives the strain tensor from the covariant components of the metric tensor. When we now define quantities

$$\zeta^{ij} = \hat{g}^{ij} - g^{ij}, \tag{2.8}$$

one might expect that these are the contravariant strain components

$$\gamma^{ij} = \gamma_{kl}\, g^{ik} g^{jl}. \tag{2.9}$$

This, however, is not the case, as we may easily see. The components $\hat{g}_{ij}$ and $\hat{g}^{ij}$ of the metric tensor of the deformed medium are, of course, related by the formula

$$\hat{g}^{ij} \hat{g}_{jk} = \hat{\delta}^{i}_{k} = \delta^{i}_{k}\,,$$

and when we use (2.5) and (2.8), we have

$$(g^{ij} + \zeta^{ij})(g_{jk} + \gamma_{jk}) = \delta^{i}_{k} + \gamma^{i}_{k} + \zeta^{i}_{k} + \zeta^{ij}\gamma_{jk} = \delta^{i}_{k}\,.$$

The last term in the second member is quadratic in strainlike quantities. When we neglect it, we find that for small strains

$$\zeta^i_k = -\gamma^i_k,$$

hence

$$\zeta^{ij} = -\gamma^{ij}$$

and

$$-\gamma^{ij} = \hat{g}^{ij} - g^{ij}. \tag{2.10}$$

The deformation of a body is completely described when, for each of its points, we know the displacement vector **u**, which extends from the position before deformation to that occupied by the same material point after deformation. It must, therefore, be possible to express $\gamma_{ij}$ in terms of **u**. This relation we shall now derive. In doing so, we shall have to differentiate a vector $\mathbf{u} = \mathbf{g}^i u_i$. Since we have not yet learned how to differentiate a base vector—a long story, which cannot be told in passing (see p. 66)—we will restrict ourselves to rectilinear coordinate systems $x^i$, in which the base vectors are constant. Then we have simply

$$d\mathbf{u} = \mathbf{g}^i \frac{\partial u_i}{\partial x^j} dx^j. \tag{2.11}$$

For the partial derivative of the vector component we introduce a new notation and write

$$\frac{\partial u_i}{\partial x^j} = u_{i,j}. \tag{2.12}$$

This comma notation will henceforth be used for all derivatives with respect to coordinates. When we introduce it into (2.11), this equation reads

$$d\mathbf{u} = \mathbf{g}^i u_{i,j}\, dx^j \tag{2.13}$$

with the summation convention applied to $i$ and $j$. However, we must keep in mind that (2.12) does not imply that $u_{i,j}$ is a tensor. If it were, we would have to prove it, but in fact it is not.

Now let us consider two adjacent material points $A$ and $B$ in the undeformed body, Figure 2.1, which are the end points of a line element vector $d\mathbf{s}$. During the deformation, $A$ undergoes the displacement **u** and moves to $\hat{A}$, while $B$ experiences a slightly different displacement $\mathbf{u} + d\mathbf{u}$ when moving to $\hat{B}$. From Figure 2.1 we read the simple vector equation

$$\mathbf{u} + d\hat{\mathbf{s}} = d\mathbf{s} + \mathbf{u} + d\mathbf{u},$$

and from this and (2.13) we have

$$d\hat{\mathbf{s}} = d\mathbf{s} + d\mathbf{u} = d\mathbf{s} + \mathbf{g}^i u_{i,j}\, dx^j. \tag{2.14}$$

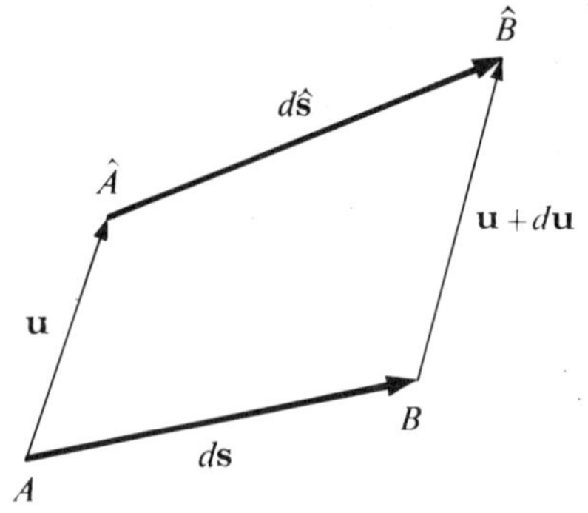


FIGURE 2.1 *Displacement and strain.*

We may now write the square of the deformed line element:

$$d\hat{\mathbf{s}} \cdot d\hat{\mathbf{s}} = (\mathbf{g}_i\, dx^i + \mathbf{g}^k u_{k,i}\, dx^i) \cdot (\mathbf{g}_j\, dx^j + \mathbf{g}^l u_{l,j}\, dx^j).$$

Since all indices are dummies, they have been chosen so that the final result looks best. When we multiply the two factors term by term and switch the notation for some dummy pairs, we obtain

$$d\hat{\mathbf{s}} \cdot d\hat{\mathbf{s}} = (g_{ij} + 2\mathbf{g}_i \cdot \mathbf{g}^l u_{l,j} + g^{kl} u_{k,i} u_{l,j})\, dx^i\, dx^j.$$

With the help of (2.4) and (2.5) this may be written

$$\gamma_{ij}\, dx^i\, dx^j = (2u_{i,j} + g^{kl} u_{k,i} u_{l,j})\, dx^i\, dx^j.$$

In this equation both $dx^i$ and $dx^j$ represent, in different sums, the three components of the same line element vector $d\mathbf{s}$. Since this vector can be chosen arbitrarily we may apply an extension of the technique explained on page 15. We first choose $d\mathbf{s} = \mathbf{g}_1\, dx^1$ and prove that the factors of $dx^1\, dx^1$ on both sides are equal, and then we do the same for $i = j = 2$ and $= 3$, but then the procedure differs because $u_{i,j} \neq u_{j,i}$. When we choose a line element $d\mathbf{s} = \mathbf{g}_1\, dx^1 + \mathbf{g}_2\, dx^2$ and cancel the terms which have already been recognized to be the same on both sides, we are left with the statement that

$$\gamma_{12} + \gamma_{21} = 2(u_{1,2} + u_{2,1}) + g^{kl}(u_{k,1} u_{l,2} + u_{k,2}\, u_{l,1}).$$

In the last term we make use of the fact that $g^{kl}$ is a constant and include it in one of the derivatives and write

$$(g^{kl} u_k)_{,1} u_{l,2} + (g^{kl} u_l)_{,1} u_{k,2} = u^l{}_{,1} u_{l,2} + u^k{}_{,1} u_{k,2} = 2u^k{}_{,1} u_{k,2}\,,$$

and since $\gamma_{12} = \gamma_{21}$, we finally have

$$\gamma_{12} = u_{1,2} + u_{2,1} + u^k{}_{,1} u_{k,2}\,.$$

Proceeding to line elements with $dx^1 = 0$ or $dx^2 = 0$, we prove that the relation

$$\gamma_{ij} = u_{i,j} + u_{j,i} + u^k{}_{,i} u_{k,j} \tag{2.15}$$

holds for all $i$, $j$.

The last term on the right-hand side of this equation is quadratic in the displacement and therefore negligible when the displacement is sufficiently small. Equation (2.15) then reduces to

$$\gamma_{ij} = u_{i,j} + u_{j,i} \,. \tag{2.16}$$

The statements made in (2.15) and (2.16) are independent of the forces acting and of the elastic or inelastic character of the material. They are concerned with the geometry of the motion which leads from the undeformed to the deformed position and are known as the *kinematic relations*.

On page 24 we called the $\gamma_{ij}$ the components of the strain tensor. This imposes the obligation of demonstrating that they are identical with the quantities commonly called strain, or at least closely related to them. For this purpose, we specialize (2.15) and (2.16) to cartesian coordinates, writing

$$dx^1 = dx, \qquad dx^2 = dy, \qquad dx^3 = dz,$$

and

$$u_1 = u, \qquad u_2 = v, \qquad u_3 = w.$$

For $i = j = 1$, (2.16) then reads

$$\gamma_{11} = 2\frac{\partial u}{\partial x}$$

and for $i = 1, j = 2$:

$$\gamma_{12} = \frac{\partial u}{\partial y} + \frac{\partial v}{\partial x}.$$

The equations are identical with the kinematic relations found in books on the theory of elasticity, if one sets

$$\gamma_{11} = 2\varepsilon_x, \qquad \gamma_{12} = \gamma_{xy}.$$

When the nonlinear equation (2.15) is subjected to the same treatment, it yields

$$\gamma_{11} = 2\frac{\partial u}{\partial x} + \left(\frac{\partial u}{\partial x}\right)^2 + \left(\frac{\partial v}{\partial y}\right)^2 + \left(\frac{\partial w}{\partial z}\right)^2,$$

$$\gamma_{12} = \frac{\partial u}{\partial y} + \frac{\partial v}{\partial x} + \frac{\partial u}{\partial x}\frac{\partial u}{\partial y} + \frac{\partial v}{\partial x}\frac{\partial v}{\partial y} + \frac{\partial w}{\partial x}\frac{\partial w}{\partial y}.$$

Also these expressions are equal to $2\varepsilon_x$ and $\gamma_{xy}$, respectively, if one adopts a strain definition based on the change of the *square* of the line element. In this case they are not second-order approximations, but are exact for any amount of deformation. This is not so if the strain $\varepsilon_x$ is defined as the increment of length divided by the original length (*linear strain*).

In the future, as a measure of the strain, we shall prefer to use the tensor

$$\varepsilon_{ij} = \tfrac{1}{2}\gamma_{ij} \tag{2.17}$$

mainly because products of the $\varepsilon_{ij}$ with the stresses are the work done during the deformation. For later reference we rewrite the kinematic relations in this notation:

small displacements (linearized formula),

$$\varepsilon_{ij} = \tfrac{1}{2}(u_{i,j} + u_{j,i}); \tag{2.18}$$

large displacements (exact formula),

$$\varepsilon_{ij} = \tfrac{1}{2}(u_{i,j} + u_{j,i} + u^k{}_{,i}u_{k,j}). \tag{2.19}$$

Since quantities like $u_{i,j}$ are not components of a tensor, these equations are not tensor equations, but, on page 85, we shall see how one can use them to derive tensor equations.

*Problem*

2.1. Use the general kinematic relations (2.18) and (2.19) to derive kinematic relations in skew rectilinear coordinates.

**References**

The subject of this chapter is treated in all books on the theory of elasticity. The kinematic relation (2.16) in cartesian component form is found in Love [19, p. 38] or in Timoshenko–Goodier [33, p. 7].

CHAPTER 3

# The Cross Product

AFTER HAVING STUDIED STRAIN, we should next study stress. Since a stress is a force acting on a certain area element, we have to see how such an area element can be expressed in tensor form. Before we can do this, we need some preparation, which is the purpose of this chapter.

## 3.1. Permutation Tensor

We define quantities $e_{ijk} = e^{ijk}$, the *permutation symbols*, by the following rules:

$e_{ijk} = +1$ if $i, j, k = 1, 2, 3$, or an even permutation of this sequence (that is, 2, 3, 1 or 3, 1, 2);

$e_{ijk} = -1$ if $i, j, k$ is an odd permutation of 1, 2, 3 (that is, 3, 2, 1 or 2, 1, 3 or 1, 3, 2);

$e_{ijk} = 0$ if any two of the subscripts or all three are equal (for example, 1, 1, 3 or 2, 3, 2).

We call these sequences $i, j, k$ cyclic, anticyclic, and acyclic, respectively. There are altogether 27 possible sequences, which are represented in Figure 3.1 as the points of the cubic grid. In this figure the six combinations for which $e_{ijk} \neq 0$ have been marked by heavy dots.

The permutation symbols can be used to expand a third-order determinant. Let

$$a = \begin{vmatrix} a_1^1 & a_2^1 & a_3^1 \\ a_1^2 & a_2^2 & a_3^2 \\ a_1^3 & a_2^3 & a_3^3 \end{vmatrix} \tag{3.1}$$

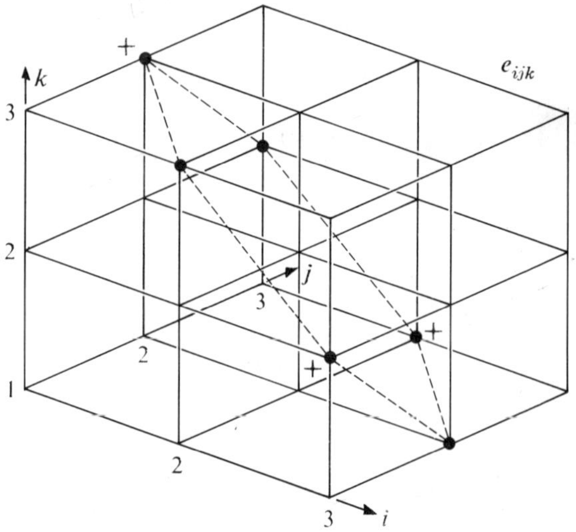


FIGURE 3.1 *Permutation symbol $e_{ijk}$.*

Then one definition of $a$ is that it is the sum of all the products $\pm a_1^i a_2^j a_3^k$ which satisfy the condition that $i, j, k$ are all different, the plus sign to be used when the sequence $i, j, k$ is cyclic and the minus sign when it is anticyclic. This can be expressed in the form

$$a = a_1^i a_2^j a_3^k e_{ijk} \tag{3.2a}$$

and in the alternate form

$$a = a_l^1 a_m^2 a_n^3 e^{lmn}. \tag{3.3}$$

If we make any permutation among the subscripts in (3.2a), e.g. if we write $a_2^i a_1^j a_3^k e_{ijk}$, this amounts to interchanging two columns of the determinant and the result will be $-a$. A further permutation would change the sign again and lead back to $+a$. This can be expressed in the form

$$ae_{lmn} = a_l^i a_m^j a_n^k e_{ijk} \tag{3.4a}$$

and similarly we have

$$ae^{ijk} = a_l^i a_m^j a_n^k e^{lmn}. \tag{3.5}$$

When we replace the determinant $a$ of (3.1) by the more general one

$$A = \begin{vmatrix} a_r^o & a_s^o & a_t^o \\ a_r^p & a_s^p & a_t^p \\ a_r^q & a_s^q & a_t^q \end{vmatrix},$$

we have $A = a$ if $o, p, q = r, s, t = 1, 2, 3$ and any permutation in one of these sequences changes the sign so that

$$A = ae^{opq}e_{rst}. \tag{3.6}$$

Equations (3.2a) through (3.5) can easily be extended to other positions of the indices. With a corresponding change in the definition of $a$ we have

$$a = |a^{pq}| = a^{i1}a^{j2}a^{k3}e_{ijk} = a^{1l}a^{2m}a^{3n}e_{lmn}, \tag{3.2b}$$

$$ae^{lmn} = a^{il}a^{jm}a^{kn}e_{ijk} \tag{3.4b}$$

and

$$a = |a_{pq}| = a_{i1}a_{j2}a_{k3}e^{ijk} = a_{1l}a_{2m}a_{3n}e^{lmn}, \tag{3.2c}$$

$$ae_{lmn} = a_{il}a_{jm}a_{kn}e^{ijk}. \tag{3.4c}$$

We use (3.2a) and (3.4a) to prove the following theorem: Given two square matrices $A$ and $B$. The determinant of their product $C = AB$ equals the product of the determinants of $A$ and $B$. The proof will be restricted to $3 \times 3$ matrices; it can easily be extended to larger and smaller ones by introducing permutation symbols with the proper number of indices.

Let $A = [a_i{}^j]$ and $B = [b_j{}^k]$. Their product $C = [c_i{}^k]$ has the elements

$$c_i{}^k = a_i{}^j b_j{}^k.$$

The determinants of $A$ and $B$ are

$$\det A = a = a_1{}^l a_2{}^m a_3{}^n e_{lmn},$$

$$\det B = b \quad \text{with} \quad e_{lmn}\, b = b_l{}^i b_m{}^j b_n{}^k e_{ijk}.$$

Their product is

$$ab = a_1{}^l a_2{}^m a_3{}^n e_{lmn}\, b = a_1{}^l b_l{}^i a_2{}^m b_m{}^j a_3{}^n b_n{}^k e_{ijk} = c_1{}^i c_2{}^j c_3{}^k e_{ijk} = c, \tag{3.7}$$

which proves the theorem.

For the transformation of vectors and tensors between two reference frames of base vectors we used the quantities $\beta^i_{i'}$ and $\beta^{i'}_i$. Because of (1.32), the matrices formed by these two sets of coefficients are reciprocals of each other and then (3.7) states that their determinants are also reciprocals:

$$|\beta^i_{i'}| = \Delta, \qquad |\beta^{i'}_i| = \frac{1}{\Delta}. \tag{3.8}$$

Quite similarly we may arrange the nine components $g_{ij}$ of the metric tensor in a square matrix and then calculate its determinant

$$g = |g_{ij}| = \begin{vmatrix} g_{11} & g_{12} & g_{13} \\ g_{21} & g_{22} & g_{23} \\ g_{31} & g_{32} & g_{33} \end{vmatrix} \tag{3.9}$$

From (1.26) it follows that

$$|g_{ij}g^{jk}| = |\delta_i^k| = 1$$

and then from (3.7) that

$$|g_{ij}|\,|g^{jk}| = 1,$$

hence

$$|g^{jk}| = \frac{1}{g}. \tag{3.10}$$

When we transform the $g_{ij}$ to another reference frame, we have

$$g_{i'j'} = g_{ij}\beta^i_{i'}\beta^j_{j'}$$

and hence by repeated applications of (3.7):

$$g' = |g_{i'j'}| = |g_{ij}\beta^i_{i'}|\,|\beta^j_{j'}| = |g_{ij}|\,|\beta^i_{i'}|\,|\beta^j_{j'}| = g\Delta\Delta. \tag{3.11}$$

This equation describes the transformation of the determinant $g$ from one reference frame to another. Although $g$ is a scalar (a tensor of order zero), it is not the same in both frames. Such quantities are called pseudoscalars, and there are pseudoscalars of different classes, depending on the positive or negative power of the transformation determinant appearing in equations like (3.11). We shall occasionally meet with such pseudoscalars, but in this book an effort has been made to avoid them, because they make the theory more complicated without being very helpful.

Thus far, the permutation symbols $e_{ijk}$ have not been attached to any coordinate system. We now attach them to a cartesian system $x^{i'}$ and write

$$e_{i'j'k'} = \epsilon_{i'j'k'}.$$

We ask what happens when we transform these quantities as tensor components to another coordinate system $x^i$. According to the general transformation rule, stated on page 19, we have

$$\epsilon_{ijk} = \epsilon_{i'j'k'}\beta^{i'}_i\beta^{j'}_j\beta^{k'}_k = e_{i'j'k'}\beta^{i'}_i\beta^{j'}_j\beta^{k'}_k. \tag{3.12}$$

For $i, j, k = 1, 2, 3$ this is (3.2a), that is the expansion formula of the determinant $1/\Delta$. Since in the cartesian system

$$g' = |\delta_{i'j'}| = 1,$$

(3.11) shows that in our case

$$\frac{1}{\Delta} = \sqrt{g}$$

and, hence,

$$\left.\begin{aligned} \epsilon_{ijk} &= +\sqrt{g} \quad \text{if} \quad i, j, k \text{ is a cyclic sequence;} \\ \epsilon_{ijk} &= -\sqrt{g} \quad \text{if} \quad i, j, k \text{ is an anticyclic sequence;} \\ \epsilon_{ijk} &= \quad 0 \quad \text{if} \quad i, j, k \text{ is an acyclic sequence.} \end{aligned}\right\} \tag{3.13}$$

The quantities $\epsilon_{ijk}$, being derived by a tensor transformation, are the covariant components of a third-order tensor, the *permutation tensor*.† Using the metric tensor $g^{ij}$, one may form components of other variances. The only set of interest are the triple-contravariant ones:

$$\epsilon^{lmn} = \epsilon_{ijk}\, g^{il} g^{jm} g^{kn}. \tag{3.14}$$

To find their numerical values, we start from the determinant of the $g^{ij}$ and apply to it the expansion formula (3.4b):

$$\frac{1}{g}\, e^{lmn} = \begin{vmatrix} g^{l1} & g^{l2} & g^{l3} \\ g^{m1} & g^{m2} & g^{m3} \\ g^{n1} & g^{n2} & g^{n3} \end{vmatrix} = g^{li} g^{mj} g^{nk} e_{ijk} = g^{li} g^{mj} g^{nk} \frac{1}{\sqrt{g}}\, \epsilon_{ijk}.$$

Because of (3.14) this is $\epsilon^{lmn}/\sqrt{g}$ and, hence,

$$\epsilon^{lmn} = \frac{1}{\sqrt{g}}\, e^{lmn}, \tag{3.15}$$

i.e. $\epsilon^{lmn}$ has the values $\pm 1/\sqrt{g}$ or 0 depending upon whether $l, m, n$ is a cyclic, an anticyclic, or an acyclic sequence.

From (3.13) it follows that $\epsilon_{ijk}$ changes its sign when any two of its indices are interchanged, and the same is true for $\epsilon^{ijk}$. In other words: the permutation tensor is antimetric with respect to any pair of indices. Therefore, its contracted product with any symmetric tensor $t^{ij} = t^{ji}$ is

$$\epsilon_{ijk}\, t^{ij} = 0 \qquad \text{and} \qquad \epsilon^{ijk} t_{ij} = 0. \tag{3.16}$$

To prove the formulas, let, for example, $k = 1$; then $i, j = 2, 3$ or $3, 2$ and

$$\epsilon_{ijk}\, t^{ij} = \epsilon_{231} t^{23} + \epsilon_{321}\, t^{32} = (\epsilon_{231} + \epsilon_{321}) t^{23},$$

and this vanishes because of (3.13).

As $\epsilon_{ijk}$ and $\epsilon^{ijk}$ have tensor character, while the permutation symbols $e_{ijk}$ and $e^{ijk}$ do not, it is useful to know that some of the determinant expansion formulas developed on page 30 can be written in terms of the $\epsilon$. Multiplying (3.4a) on both sides by $\sqrt{g}$ and dividing (3.5) by the same quantity, we see that

† Here we are facing one of the cases (not uncommon nowadays) where established notations of two previously separated fields come to a clash. We shall use the "English" character $\epsilon$ for the permutation tensor and the "continental" character $\varepsilon$ for the strain.

$$a\epsilon_{lmn} = a^i_l a^j_m a^k_n \epsilon_{ijk}, \tag{3.17}$$

$$a\epsilon^{ijk} = a^i_l a^j_m a^k_n \epsilon^{lmn}. \tag{3.18}$$

The nine quantities $\delta^i_j$ are the elements of a unit matrix, and their determinant equals unity:

$$\begin{vmatrix} \delta^1_1 & \delta^1_2 & \delta^1_3 \\ \delta^2_1 & \delta^2_2 & \delta^2_3 \\ \delta^3_1 & \delta^3_2 & \delta^3_3 \end{vmatrix} = 1.$$

To this determinant we apply (3.6) and, realizing that, because of (3.13) and (3.15),

$$e^{ijk}e_{lmn} = \epsilon^{ijk}\epsilon_{lmn},$$

we find that

$$\begin{vmatrix} \delta^i_l & \delta^i_m & \delta^i_n \\ \delta^j_l & \delta^j_m & \delta^j_n \\ \delta^k_l & \delta^k_m & \delta^k_n \end{vmatrix} = \epsilon^{ijk}\epsilon_{lmn}.$$

Expansion of the determinant yields the formula

$$\epsilon^{ijk}\epsilon_{lmn} = \delta^i_l\delta^j_m\delta^k_n - \delta^i_l\delta^j_n\delta^k_m + \delta^i_n\delta^j_l\delta^k_m - \delta^i_n\delta^j_m\delta^k_l + \delta^i_m\delta^j_n\delta^k_l - \delta^i_m\delta^j_l\delta^k_n. \tag{3.19}$$

Through the process of contraction we derive from it the following useful relations:

$$\begin{aligned} \epsilon^{ijk}\epsilon_{imn} &= \delta^i_i(\delta^j_m\delta^k_n - \delta^j_n\delta^k_m) + \delta^i_n(\delta^j_i\delta^k_m - \delta^j_m\delta^k_i) + \delta^i_m(\delta^j_n\delta^k_i - \delta^j_i\delta^k_n) \\ &= \delta^j_m\delta^k_n - \delta^j_n\delta^k_m, \end{aligned} \tag{3.20}$$

$$\epsilon^{ijk}\epsilon_{ijn} = 3\delta^k_n - \delta^j_n\delta^k_j = 2\delta^k_n, \tag{3.21}$$

$$\epsilon^{ijk}\epsilon_{ijk} = 2\delta^k_k = 6. \tag{3.22}$$

With the results just obtained, we may bring the determinant expansion formula (3.17) into another, very useful form. We multiply on both sides by $\epsilon^{lmn}$ and find that

$$a\epsilon_{lmn}\epsilon^{lmn} = 6a = a^i_l a^j_m a^k_n \epsilon_{ijk}\epsilon^{lmn}. \tag{3.23}$$

In an antimetric tensor $b_{ij}$ (see p. 18) the components with $i = j$ are zero and the others are interdependent in pairs by the relation $b_{ij} = -b_{ji}$. Therefore, such a tensor has just as many independent components as a vector. The permutation tensor permits the association of a definite vector $u^k$ with every antimetric tensor $b_{ij}$ through the relations

$$u^k = b_{ij}\epsilon^{ijk}, \qquad u_k = b^{ij}\epsilon_{ijk}. \tag{3.24}$$

One of the components of this vector is

$$u^1 = (b_{23} - b_{32})\epsilon^{231}.$$

Equation (3.24) may also be used with any unsymmetric tensor thereby isolating from it the antimetric part.

Thus far we have always dealt with three-dimensional space, using coordinates $x^i$, $i = 1, 2, 3$. Many problems in mechanics are formulated in only two dimensions. For example, plane stress systems and plane flow fields are described in terms of plane coordinates $x^1$, $x^2$, and in the theory of shells we use two coordinates $x^1$, $x^2$ on a curved surface. It has become general usage to emphasize the difference between equations valid in two and in three dimensions by observation of the following

RANGE CONVENTION: All Latin indices have the range $i, j, k, l, m, \ldots = 1, 2, 3$; while all Greek indices have the range $\alpha, \beta, \gamma, \delta, \ldots = 1, 2$.

This applies to free indices and to dummies. The vector equation $a_i = b_i$ represents three component equations, while $a_\alpha = b_\alpha$ represents two. The sum $a^i b_i$ has three terms, but the sum $a^\alpha b_\alpha$ has two terms.

We may consider any two-dimensional space as a subspace of the general three-dimensional space, adding to its coordinates $x^\alpha$ a third coordinate $x^3$ with a unit base vector $\mathbf{g}_3 = \mathbf{i}_3$ normal to the vectors $g_\alpha$. We have then $g_{\alpha 3} = \mathbf{g}_\alpha \cdot \mathbf{g}_3 = 0$ and $g_{33} = 1$, and the determinant of the metric tensor is

$$g = \begin{vmatrix} g_{11} & g_{12} & 0 \\ g_{21} & g_{22} & 0 \\ 0 & 0 & 1 \end{vmatrix} = \begin{vmatrix} g_{11} & g_{12} \\ g_{21} & g_{22} \end{vmatrix}. \tag{3.25}$$

In every nonvanishing component $\epsilon_{ijk}$ of the permutation tensor one of the subscripts must be a 3, and we can, by cyclic permutation, always arrange it so that $k = 3$. Then $i$ and $j$ are restricted to the range 1, 2 and we may introduce a two-dimensional permutation tensor $\epsilon_{\alpha\beta}$, writing

$$\epsilon_{ij3} = \epsilon_{\alpha\beta 3} = \epsilon_{\alpha\beta} \tag{3.26a}$$

with

$$\epsilon_{11} = \epsilon_{22} = 0, \qquad \epsilon_{12} = \sqrt{g} = -\epsilon_{21}. \tag{3.27a}$$

The contravariant components are

$$\epsilon^{ij3} = \epsilon^{\alpha\beta 3} = \epsilon^{\alpha\beta} \tag{3.26b}$$

with

$$\epsilon^{11} = \epsilon^{22} = 0, \qquad \epsilon^{12} = \frac{1}{\sqrt{g}} = -\epsilon^{21}. \tag{3.27b}$$

At the proper places we shall make much use of these quantities. At present, we list a few formulas for later reference, leaving the proofs to the reader (see the exercises on page 43).

For a second-order determinant

$$a = \begin{vmatrix} a_1^1 & a_2^1 \\ a_1^2 & a_2^2 \end{vmatrix}$$

the following formulas hold:

$$a\epsilon_{\gamma\delta} = a_\gamma^\alpha\, a_\delta^\beta \epsilon_{\alpha\beta}\,, \qquad a\epsilon^{\alpha\beta} = a_\gamma^\alpha a_\delta^\beta \epsilon^{\gamma\delta}, \tag{3.28a, b}$$

$$2a = a_\gamma^\alpha\, a_\delta^\beta \varepsilon_{\alpha\beta}\, \epsilon^{\gamma\delta}. \tag{3.28c}$$

Products of the permutation tensors have the following values:

$$\epsilon^{\alpha\beta}\epsilon_{\gamma\delta} = \delta_\gamma^\alpha\, \delta_\delta^\beta - \delta_\delta^\alpha\, \delta_\gamma^\beta\,, \tag{3.29a}$$

$$\epsilon^{\alpha\beta}\epsilon_{\alpha\gamma} = \delta_\gamma^\beta\,, \qquad \epsilon^{\alpha\beta}\epsilon_{\alpha\beta} = 2. \tag{3.29b, c}$$

For an unsymmetric tensor $T_{\gamma\delta}$, the difference $T_{\gamma\delta} - T_{\delta\gamma}$ does not vanish and can be written in the following form:

$$T_{\gamma\delta} - T_{\delta\gamma} = T_{\alpha\beta}(\delta_\gamma^\alpha\, \delta_\delta^\beta - \delta_\delta^\alpha\, \delta_\gamma^\beta) = T_{\alpha\beta}\, \epsilon^{\alpha\beta}\epsilon_{\gamma\delta}\,. \tag{3.30}$$

If a tensor $T_{\alpha\beta}$ is symmetric, then

$$T_{\alpha\beta}\, \epsilon^{\alpha\beta} = 0. \tag{3.31}$$

The vector $\mathbf{v} = u_\alpha\, \epsilon^{\alpha\beta}\mathbf{g}_\beta$ is normal to the vector $\mathbf{u} = u_\gamma \mathbf{g}^\gamma$. We prove this statement by showing that the dot product of the two vectors equals zero:

$$\mathbf{v} \cdot \mathbf{u} = u_\alpha\, \epsilon^{\alpha\beta}\mathbf{g}_\beta \cdot u_\gamma \mathbf{g}^\gamma = u_\alpha\, u_\gamma\, \epsilon^{\alpha\beta}\delta_\beta^\gamma = u_\alpha\, u_\beta\, \epsilon^{\alpha\beta}.$$

Since the tensor $T_{\alpha\beta} = u_\alpha\, u_\beta$ is symmetric, it follows from (3.31) that $\mathbf{v} \cdot \mathbf{u} = 0$.

## 3.2. Cross Product

In a cartesian reference frame $\mathbf{i}, \mathbf{j}, \mathbf{k}$ we introduce the cross product of two vectors by the equations

$$\mathbf{i} \times \mathbf{j} = \mathbf{k}, \qquad \mathbf{j} \times \mathbf{k} = \mathbf{i}, \qquad \mathbf{k} \times \mathbf{i} = \mathbf{j}.$$

Writing $\mathbf{g}_i$ or $\mathbf{g}^k$ for the unit vectors, we may summarize these three equations in the form

$$\mathbf{g}_i \times \mathbf{g}_j = e_{ijk}\, \mathbf{g}^k.$$

Since in the cartesian reference frame $e_{ijk} = \epsilon_{ijk}$, we may write as well

$$\mathbf{g}_i \times \mathbf{g}_j = \epsilon_{ijk}\, \mathbf{g}^k, \tag{3.32}$$

and this equation is written entirely in tensor symbols and carries over in all other reference frames. If $\mathbf{g}_1, \mathbf{g}_2, \mathbf{g}_3$ in this order are a right-handed frame, so are the vectors $\mathbf{g}_i, \mathbf{g}_j, \mathbf{g}_i \times \mathbf{g}_j$.

Instead of (3.32) we might have written

$$\mathbf{g}^i \times \mathbf{g}^j = \epsilon^{ijk}\mathbf{g}_k, \tag{3.33}$$

which is also a tensor formula and, hence, generally valid. Both $\epsilon$ symbols change sign when two of their indices are interchanged. We therefore have

$$\mathbf{g}_j \times \mathbf{g}_i = -\mathbf{g}_i \times \mathbf{g}_j, \qquad \mathbf{g}^j \times \mathbf{g}^i = -\mathbf{g}^i \times \mathbf{g}^j. \tag{3.34}$$

When visualizing the cross product of two vectors, we shall find it useful to adhere to the right-hand rule. It should, however, be kept in mind that there is nothing wrong with left-handed reference frames and that, in exceptional cases, the vectors $\mathbf{g}_i$ associated with a curvilinear coordinate system may be right-handed in some part of space and left-handed in another.

The definition of the cross product of two base vectors may easily be extended to arbitrary vectors **a** and **b** by writing

$$\mathbf{q} = \mathbf{a} \times \mathbf{b} = a^i\mathbf{g}_i \times b^j\mathbf{g}_j = a^ib^j\epsilon_{ijk}\mathbf{g}^k = q_k\mathbf{g}^k \tag{3.35a}$$

with

$$q_k = a^ib^j\epsilon_{ijk} \tag{3.35b}$$

or

$$\mathbf{q} = a_i\mathbf{g}^i \times b_j\mathbf{g}^j = a_ib_j\epsilon^{ijk}\mathbf{g}_k = q^k\mathbf{g}_k \tag{3.36a}$$

with

$$q^k = a_ib_j\epsilon^{ijk}. \tag{3.36b}$$

Also in this case the product vector is at right angles with each of the factors. We show this for **a** by forming the dot product

$$\mathbf{q} \cdot \mathbf{a} = q_k a^k = a^ib^j\epsilon_{ijk}a^k = a^ib^ja^ke_{ijk}\sqrt{g}.$$

This is the expansion formula for a determinant whose rows are the components of the vectors **a**, **b**, **a**. Since two of these rows are equal, the determinant equals zero, and this proves the orthogonality of **q** and **a**.

For our purposes it is important that the absolute value of the vector $= \mathbf{a} \times \mathbf{b}$ is equal to the area $A$ of the parallelogram whose sides are the ectors **a** and **b**. We prove this statement by proving that the components of in a cartesian reference frame $\mathbf{g}_i$ are equal to the areas of the projections f this parallelogram on the reference planes. Let

$$\mathbf{a} = a^i\mathbf{g}_i, \qquad \mathbf{b} = b^j\mathbf{g}_j, \qquad \mathbf{q} = q_k\mathbf{g}^k,$$

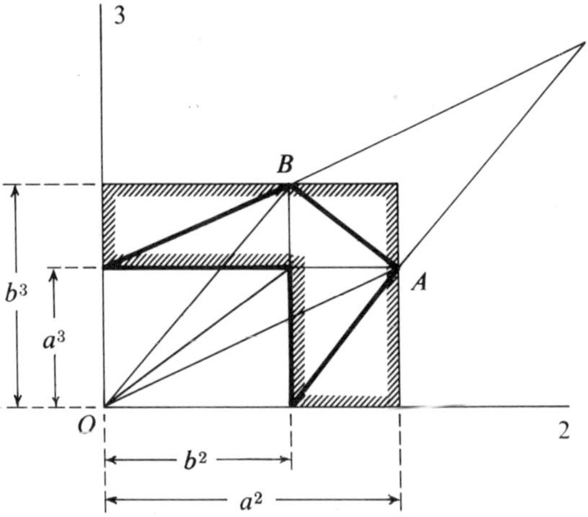


FIGURE 3.2 *Calculation of the cross product.*

then

$$q_k = a^i b^j e_{ijk}.$$

We consider now one of these components, say

$$q_1 = a^2 b^3 - a^3 b^2.$$

In the plane of the unit vectors $\mathbf{g}_2$ and $\mathbf{g}_3$ this appears as the difference of two rectangles. Figure 3.2 shows the projections $OA$ and $OB$ of the vectors $\mathbf{a}$ and $\mathbf{b}$ on this plane. The area representing $q_1$ is indicated by contour shading. Half of it is the area enclosed by the heavy line, and this is obviously equal to that of the triangle $OAB$. This, in turn, is half of the parallelogram $OACB$, which is the projection of the parallelogram area $A$ on the 2, 3 plane and a similar statement can be proved for the components $q_2$ and $q_3$.

The fact that the cross product is the vector representation of an area is of great importance in mechanics of continuous media, since we need an area element to define stresses. Before we take up this subject, we will learn a few useful facts connected with the cross product.

When we dot-multiply the cross product of two vectors with a third vector we have

$$\mathbf{a} \times \mathbf{b} \cdot \mathbf{c} = a^i b^j \epsilon_{ijk} \mathbf{g}^k \cdot c^l \mathbf{g}_l = a^i b^j c^l \epsilon_{ijk} \delta_l^k = a^i b^j c^k \epsilon_{ijk}. \tag{3.37}$$

On the other hand, from (1.2) and Figure 3.3 we have

$$\mathbf{a} \times \mathbf{b} \cdot \mathbf{c} = |\mathbf{a} \times \mathbf{b}|\ |\mathbf{c}| \cos \beta,$$

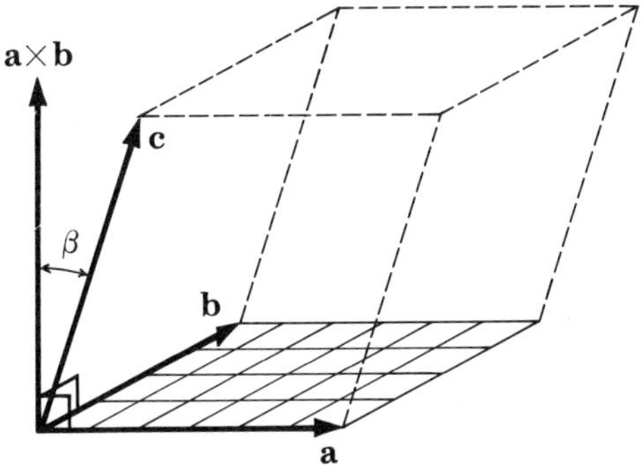


FIGURE 3.3 *The product* $\mathbf{a} \times \mathbf{b} \cdot \mathbf{c}$.

and this is the volume $V$ of the skew block outlined in the figure, which, hence, can be written in the form

$$V = a^i b^j c^k \epsilon_{ijk} = \mathbf{a} \times \mathbf{b} \cdot \mathbf{c}. \tag{3.38}$$

In the double product $\mathbf{a} \times \mathbf{b} \cdot \mathbf{c}$ it is not customary to indicate by parentheses that the cross-multiplication has to be done first, because there is no other choice. The dot product $\mathbf{b} \cdot \mathbf{c}$ is a scalar, and the cross product between this scalar and the vector $\mathbf{a}$ is meaningless.

When we interchange any two factors in the product $\mathbf{a} \times \mathbf{b} \cdot \mathbf{c}$, we must interchange the corresponding subscripts in $\epsilon_{ijk}$, e.g.

$$\mathbf{a} \times \mathbf{c} \cdot \mathbf{b} = a^i c^k b^j \epsilon_{ikj} = -a^i c^k b^j \epsilon_{ijk} = -V.$$

From this we see that a cyclic interchange of factors does not change the result, while an anticyclic sequence of factors produces the opposite sign:

$$\begin{aligned} \mathbf{a} \times \mathbf{b} \cdot \mathbf{c} &= \mathbf{c} \times \mathbf{a} \cdot \mathbf{b} = \mathbf{b} \times \mathbf{c} \cdot \mathbf{a} \\ &= -\mathbf{c} \times \mathbf{b} \cdot \mathbf{a} = -\mathbf{a} \times \mathbf{c} \cdot \mathbf{b} = -\mathbf{b} \times \mathbf{a} \cdot \mathbf{c}. \end{aligned} \tag{3.39}$$

Moreover, since the two factors of a dot product can be interchanged, we have

$$\mathbf{b} \times \mathbf{c} \cdot \mathbf{a} = \mathbf{a} \cdot \mathbf{b} \times \mathbf{c},$$

and by comparison with the preceding equation

$$\mathbf{a} \cdot \mathbf{b} \times \mathbf{c} = \mathbf{a} \times \mathbf{b} \cdot \mathbf{c}. \tag{3.40}$$

This shows that it does not matter where we place the cross and where the dot; all that matters is the sequence of the factors. The product is positive when they are so arranged that they form a right-handed system (more precisely: a system of the same "handedness" as the reference frame $\mathbf{g}_i$ on which the definition of the cross product is based).

Since, from (3.32),

$$\mathbf{g}_1 \times \mathbf{g}_2 = \epsilon_{123}\mathbf{g}^3 = \sqrt{g}\,\mathbf{g}^3,$$

we have

$$\mathbf{g}_1 \times \mathbf{g}_2 \cdot \mathbf{g}_3 = \sqrt{g}\,\mathbf{g}^3 \cdot \mathbf{g}_3 = \sqrt{g} \tag{3.41}$$

and, similarly, from (3.33)

$$\mathbf{g}^1 \times \mathbf{g}^2 \cdot \mathbf{g}^3 = \epsilon^{123}\mathbf{g}_3 \cdot \mathbf{g}^3 = \frac{1}{\sqrt{g}}. \tag{3.42}$$

This interprets $\sqrt{g}$ as the volume of a block† that has the covariant base vectors for its edges.

When we rewrite (3.35) in the form

$$d\mathbf{A} = d\mathbf{r} \times d\mathbf{s}, \qquad dA_k = dr^i\, ds^j\, \epsilon_{ijk}, \tag{3.43}$$

we are representing an area element by a vector normal to it (Figure 3.4a). On the other hand, a volume element

$$dV = d\mathbf{r} \times d\mathbf{s} \cdot d\mathbf{t} = dr^i\, ds^j\, dt^k\, \epsilon_{ijk} \tag{3.44}$$

is a scalar (Figure 3.4b). In continuum mechanics we frequently meet quantities which are defined per unit of an area or a volume. We shall now inspect a few samples of this kind.

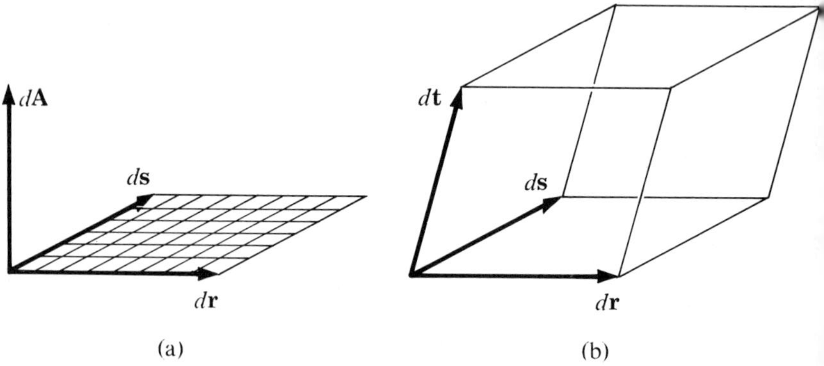


FIGURE 3.4 *Area and volume elements.*

† Commonly called a parallelepipedon. It seems time that our language adopt a word that is easier to pronounce and to spell than this monster.

The mass $dm$ of a volume element is the product of the density $\rho$ and the volume $dV$:

$$dm = \rho\, dV = \rho\, dr^i\, ds^j\, dt^k\, \epsilon_{ijk}.$$

We subject the base vectors $\mathbf{g}_i$ to a transformation

$$\mathbf{g}_i = \beta_i^{l'} \mathbf{g}_{l'}, \qquad dr^i = \beta_{l'}^i\, dr^{l'}$$

and have

$$\begin{aligned} dm &= \rho \beta_{l'}^i \beta_{m'}^j \beta_{n'}^k\, dr^{l'}\, ds^{m'}\, dt^{n'} \beta_i^{p'} \beta_j^{q'} \beta_k^{r'} \epsilon_{p'q'r'} \\ &= \rho \delta_{l'}^{p'} \delta_{m'}^{q'} \delta_{n'}^{r'}\, dr^{l'}\, ds^{m'}\, dt^{n'} \epsilon_{p'q'r'} \\ &= \rho\, dr^{p'}\, ds^{q'}\, dt^{r'}\, \epsilon_{p'q'r'}. \end{aligned}$$

The volume element appears in the same form, as was to be expected, and $\rho$ is a scalar, the same in all coordinate systems.

The same is true for the pressure $p$ of a fluid or a gas. Its product with an area of element $d\mathbf{A}$, on which it acts, is a force, and this force is normal to the area, i.e. it has the direction of $d\mathbf{A}$. If we arrange things so that $d\mathbf{A}$ is the outer normal, the force is

$$d\mathbf{F} = -p\, d\mathbf{A}$$

with components

$$dF_k = -p\, dA_k = -p\, dr^i\, ds^j\, \epsilon_{ijk}.$$

Again, when we transform to another reference frame, $dr^i\, ds^j\, \epsilon_{ijk} \mathbf{g}^k$ transforms into an expression of the same form in the new frame and the scalar $p$ remains unchanged.

The weight $d\mathbf{W}$ of a volume element $dV$ is a vector, the product of the volume and the specific weight

$$\boldsymbol{\gamma} = \gamma^l \mathbf{g}_l,$$

hence

$$d\mathbf{W} = \gamma^l\, \mathbf{g}_l\, dV = \gamma^l\, \mathbf{g}_l\, dr^i\, ds^j\, dt^k\, \epsilon_{ijk}.$$

In another reference frame $\mathbf{g}_{i'}$ the same weight of the same volume is

$$d\mathbf{W} = \gamma^{l'} \mathbf{g}_{l'}\, dr^{i'}\, ds^{j'}\, dt^{k'} \epsilon_{i'j'k'},$$

and $\gamma^l$ transforms like any contravariant vector component. The same is true for the forces $X^l$ per unit volume (or per unit mass) which occur in the equilibrium conditions of continuum mechanics.

Thus far, the two factors of our cross products had always the dimension of a length. On page 2 we have introduced the dot product as the work product of a length and a force. There exists a second product of a force and a distance, the moment of a force **P** at the lever arm **r**, and this is the cross product of the two quantities:

$$\mathbf{M} = \mathbf{r} \times \mathbf{P}. \tag{3.45}$$

When all three vectors are represented in components with respect to a reference frame, (3.35b) gives the covariant components of the moment as

$$M_k = r^i P^j \epsilon_{ijk}. \tag{3.46}$$

In this book we shall mostly restrict our interest to the actual space of three dimensions and to a two-dimensional subspace. In connection with the cross product, however, it is interesting to have a brief look at its possible extension to four dimensions.

Let us first go back one step and restrict the vectors **a** and **b** of (3.35) to two dimensions:

$$\mathbf{a} = a^\alpha \mathbf{g}_\alpha, \qquad \mathbf{b} = b^\beta \mathbf{g}_\beta.$$

Then $k$ in (3.35) must necessarily equal 3 and the cross product **q** has only one component

$$q_3 = a^\alpha b^\beta \epsilon_{\alpha\beta}.$$

In any two-dimensional transformation, which changes the base vectors $\mathbf{g}_\alpha$ into new vectors $\mathbf{g}_{\gamma'}$, the unit vector $\mathbf{g}^3 = \mathbf{g}_3$ normal to $\mathbf{g}_\alpha$ and $\mathbf{g}_{\gamma'}$ is not changed and neither is $q_3$; that is, the cross product is a scalar.

In four dimensions we use indices $I, J, K, L \ldots$ with the range 1, 2, 3, 4 and define a permutation symbol $e_{IJKL}$ by the following rules:

$e_{IJKL} = \phantom{-}1$ if $I, J, K, L = 1, 2, 3, 4$ or an even permutation of this sequence,
$e_{IJKL} = -1$ if $I, J, K, L$ is an odd permutation of 1, 2, 3, 4,
$e_{IJKL} = \phantom{-}0$ if not all four subscripts are different.

Associating $e_{IJKL}$ with a cartesian reference frame and then transforming to a noncartesian frame produces the permutation tensor $\epsilon_{IJKL}$ as described on page 32.

We now have the choice of two courses to take. The first one is to define the components of the cross product of two vectors

$$\mathbf{a} = a^I \mathbf{g}_I \qquad \text{and} \qquad \mathbf{b} = b^J \mathbf{g}_J$$

as

$$q_{KL} = a^I b^J \epsilon_{IJKL}. \tag{3.47}$$

This cross product is a second-order tensor. As in two or three dimensions, its components change sign when the sequence of the factors is inverted:

$$\mathbf{b} \times \mathbf{a} = -\,\mathbf{a} \times \mathbf{b},$$

and in addition, the tensor $q_{KL}$ is antimetric:

$$q_{KL} = -q_{LK}.$$

The second choice of a generalization of the cross product consists in introducing a third vector $\mathbf{c} = c^K \mathbf{g}_K$ and in defining a triple cross product

$$\mathbf{q} = \mathbf{a} \times \mathbf{b} \times \mathbf{c} = q_L \mathbf{g}^L$$

with

$$q_L = a^I b^J c^K \epsilon_{IJKL}, \tag{3.48}$$

and this product is a four-dimensional vector. As in its three-dimensional counterpart, the interchange of two of its factors changes the sign of the result.

Each of these extensions of the cross product may be extended further to spaces of any number of dimensions. It may be mentioned in passing, that the extension of the dot product is trivial. All that is needed is to extend in (1.13) the range of the indices.

## *Problems*

3.1. Use tensor notation to prove the following vector formulas:
(a) $(\mathbf{a} \times \mathbf{b}) \cdot (\mathbf{c} \times \mathbf{d}) = (\mathbf{a} \cdot \mathbf{c})(\mathbf{b} \cdot \mathbf{d}) - (\mathbf{a} \cdot \mathbf{d})(\mathbf{b} \cdot \mathbf{c})$,
(b) $\mathbf{a} \times (\mathbf{b} \times \mathbf{c}) = (\mathbf{a} \cdot \mathbf{c})\mathbf{b} - (\mathbf{a} \cdot \mathbf{b})\mathbf{c}$.

3.2. Derive equations (3.28) and (3.29).

CHAPTER 4

# Stress

TO DEFINE STRESS, we need an area element of arbitrary size and orientation without insisting that it have any particular shape. On the rear side of this element there is some material and on its front side there is empty space. An outer normal points into this empty space.

## 4.1. Stress Tensor

We have seen that area elements in the shape of a parallelogram can be represented by a vector $d\mathbf{A}$ with components $dA_i$. To visualize these components, we proceed as follows: We choose a plane of the desired orientation and intersect it with the base vectors $\mathbf{g}_i$ of a reference frame (Figure 4.1a). We connect the points of intersection to form a triangle $ABC$ and then move the plane parallel to itself until this triangle has the desired size and thus is the area element $d\mathbf{A}$ which we want to consider. Denoting two of its sides by $d\mathbf{r}$ and $d\mathbf{s}$ as shown, we may write

$$d\mathbf{A} = \tfrac{1}{2}\, d\mathbf{r} \times d\mathbf{s} \tag{4.1}$$

and thus represent the element by a vector in the direction of its outer normal. Since $d\mathbf{r} = d\mathbf{b} - d\mathbf{a}$ and $d\mathbf{s} = d\mathbf{c} - d\mathbf{a}$, we may write (4.1) in the form

$$d\mathbf{A} = \tfrac{1}{2}(d\mathbf{b} - d\mathbf{a}) \times (d\mathbf{c} - d\mathbf{a}) = \tfrac{1}{2}(d\mathbf{b} \times d\mathbf{c} + d\mathbf{c} \times d\mathbf{a} + d\mathbf{a} \times d\mathbf{b}).$$

The vectors

$$d\mathbf{a} = da^1\, \mathbf{g}_1, \qquad d\mathbf{b} = db^2\, \mathbf{g}_2, \qquad d\mathbf{c} = dc^3\, \mathbf{g}_3 \tag{4.2}$$

have each only one nonvanishing contravariant component, and their cross products point in the directions of the base vectors $\mathbf{g}^i$:

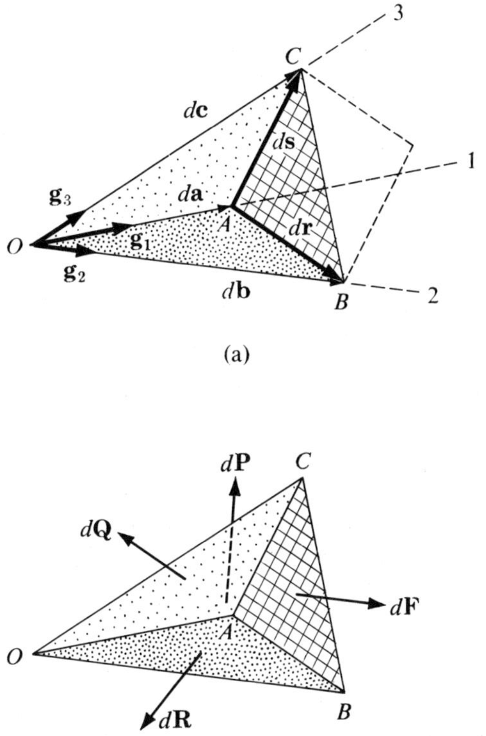


(b)

FIGURE 4.1 *Definition of stress.*

$$d\mathbf{A} = \tfrac{1}{2}(db^2\, dc^3\, \epsilon_{231}\,\mathbf{g}^1 + dc^3\, da^1\, \epsilon_{312}\,\mathbf{g}^2 + da^1\, db^2\, \epsilon_{123}\,\mathbf{g}^3) = dA_i\,\mathbf{g}^i$$

with

$$dA_1 = \tfrac{1}{2}\epsilon_{123}\, db^2\, dc^3, \qquad dA_2 = \tfrac{1}{2}\epsilon_{123}\, dc^3\, da^1, \qquad dA_3 = \tfrac{1}{2}\epsilon_{123}\, da^1\, db^2. \quad (4.3)$$

On the other hand, the triangle $OAB$ can be represented by its outer normal

$$\tfrac{1}{2}\, d\mathbf{b} \times d\mathbf{a} = \tfrac{1}{2}\, db^2\, da^1\, \epsilon_{213}\,\mathbf{g}^3 = -dA_3\,\mathbf{g}^3$$

and similar statements hold for $OBC$ and $OCA$. The four vectors representing the four sides of the tetrahedron $OABC$ add up to zero, and the *inner* normals of the three sides joined at $O$ are the covariant components of $d\mathbf{A}$.

Now let this tetrahedron be cut from some material and let forces $d\mathbf{P}$, $d\mathbf{Q}$, $d\mathbf{R}$, and $d\mathbf{F}$ act on it as shown in Figure 4.1b. Each of them is proportional to the area on which it acts and can be resolved into contravariant components

$$dP = -\sigma^{1j}\,dA_1\,g_j. \qquad dQ = -\sigma^{2j}\,dA_2\,g_j, \qquad dR = -\sigma^{3j}\,dA_3\,g_j. \quad (4.4)$$

They define nine quantities $\sigma^{ij}$, which we may understand as force components referred to the size of the areas in which they are transmitted. However, $\sigma^{1j}\,dA_1$ is not exactly a force, but only becomes one when multiplied by the base vector $\mathbf{g}_j$ (which may not be a unit vector and may not even be dimensionless), and $\sigma^{12}\mathbf{g}_2$ is not a force per unit area, because $dA_1$ is not an area unless multiplied by $\mathbf{g}^1$ (which is not a unit vector and may have dimension). The quantities $\sigma^{ij}$ come as close to the usual concept of stress as it is possible to come within the concepts of tensor mechanics.

Equilibrium of the tetrahedron in Figure 4.1b demands that the sum of all forces acting on it equal zero, whence

$$d\mathbf{F} = -\,d\mathbf{P} - d\mathbf{Q} - d\mathbf{R} = \sigma^{ij}\,dA_i\,\mathbf{g}_j = dF^j\,\mathbf{g}_j \tag{4.5}$$

with the components

$$dF^j = \sigma^{ij}\,dA_i. \tag{4.6}$$

This equation shows that the $\sigma^{ij}$ do not only describe the forces acting on three sides of the tetrahedron *OABC*, but also describe those in any arbitrary area element determined by its components $dA_i$. Since $dF^j$ and $dA_i$ are vector components, it follows that $\sigma^{ij}$ are the components of a second-order tensor, the *stress tensor*.

When we choose as the reference frame $\mathbf{g}_i$ a cartesian system of unit vectors **i**, **j**, **k**, the stresses $\sigma^{ij}$ become identical with the commonly used stresses: $\sigma^{11} = \sigma_{xx} = \sigma_x$, $\sigma^{12} = \sigma_{xy} = \tau_{xy}$, etc. Since $\tau_{xy} = \tau_{yx}$, we may ask whether a similar relation holds between $\sigma^{ij}$ and $\sigma^{ji}$. To find the answer, we study the moment equilibrium of the skew block shown in Figure 4.2. Three of its edges are arbitrary vectors $d\mathbf{a}$, $d\mathbf{b}$, $d\mathbf{c}$. The forces acting on three of the faces are

$$\begin{aligned} d\mathbf{P} &= dP^m\,\mathbf{g}_m = \sigma^{lm}\,db^j\,dc^k\,\epsilon_{ljk}\,\mathbf{g}_m, \\ d\mathbf{Q} &= dQ^m\,\mathbf{g}_m = \sigma^{lm}\,dc^k\,da^i\,\epsilon_{lki}\,\mathbf{g}_m, \\ d\mathbf{R} &= dR^m\,\mathbf{g}_m = \sigma^{lm}\,da^i\,db^j\,\epsilon_{lij}\,\mathbf{g}_m. \end{aligned}$$

There are equal forces of opposite direction on the opposite faces of the block and, together, they form three couples, which have the moments

$$\begin{aligned} d\mathbf{a} \times d\mathbf{P} &= da^i\,dP^m\,\epsilon_{imn}\,\mathbf{g}^n = \sigma^{lm}\,da^i\,db^j\,dc^k\,\epsilon_{jkl}\,\epsilon_{imn}\,\mathbf{g}^n, \\ d\mathbf{b} \times d\mathbf{Q} &= db^j\,dQ^m\,\epsilon_{jmn}\,\mathbf{g}^n = \sigma^{lm}\,da^i\,db^j\,dc^k\,\epsilon_{kil}\,\epsilon_{jmn}\,\mathbf{g}^n, \\ d\mathbf{c} \times d\mathbf{R} &= dc^k\,dR^m\,\epsilon_{kmn}\,\mathbf{g}^n = \sigma^{lm}\,da^i\,db^j\,dc^k\,\epsilon_{ijl}\,\epsilon_{kmn}\,\mathbf{g}^n. \end{aligned}$$

Equilibrium requires that the sum of these three moments be zero. This is a vector represented by its covariant components, $\mathbf{M} = M_n\mathbf{g}^n$, and each component $M_n$ must vanish:

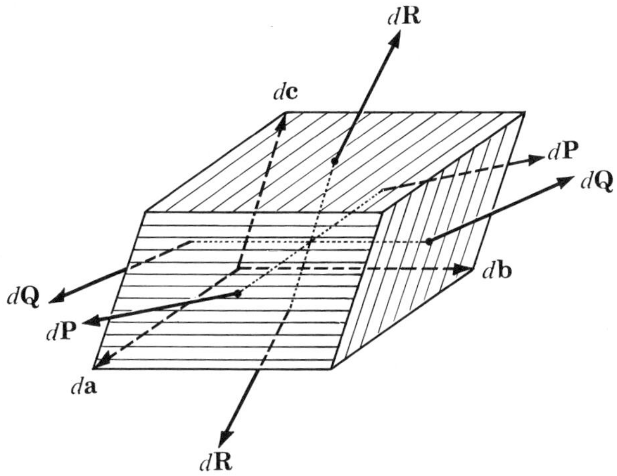


FIGURE 4.2 *Moment equilibrium of stresses.*

$$\sigma^{lm}\, da^i\, db^j\, dc^k (\epsilon_{jkl}\,\epsilon_{imn} + \epsilon_{kil}\,\epsilon_{jmn} + \epsilon_{ijl}\,\epsilon_{kmn}) = 0. \tag{4.7}$$

This equation is valid separately for each $n$, but contains five dummy indices and, hence, $3^5 = 243$ terms. Of course, most of them are zero. Still, it is worthwhile to avoid a lengthy search for the nonvanishing terms and to specialize, choosing the vectors $d\mathbf{a}$, $d\mathbf{b}$, $d\mathbf{c}$ to be those described by (4.2). Then we have always $i = 1$, $j = 2$, $k = 3$. For $n = 1$, (4.7) then reads simply

$$\sigma^{lm}(\epsilon_{23l}\,\epsilon_{1m1} + \epsilon_{31l}\,\epsilon_{2m1} + \epsilon_{12l}\,\epsilon_{3m1}) = 0.$$

The first term in the parentheses contains $\epsilon_{1m1} = 0$. In the other two only two choices are left: $l = 2$, $m = 3$ and $l = 3$, $m = 2$, respectively, and all that is left of the many sums is

$$\sigma^{23}\epsilon_{312}\,\epsilon_{231} + \sigma^{32}\epsilon_{123}\,\epsilon_{321} = 0.$$

The first three of the components of the permutation tensor have cyclic subscripts, but in the last one they are anticylic, hence

$$\sigma^{23} - \sigma^{32} = 0.$$

Choosing in (4.7) $n = 2$ or $n = 3$ yields similar statements for the other pairs of stress components and proves in general that the stress tensor is symmetric:

$$\sigma^{ij} = \sigma^{ji}. \tag{4.8}$$

Thus far, we have only considered the contravariant components of the stress tensor. We may, of course, make use of the metric tensor and introduce other types:

$$\sigma^{i}_{\cdot j} = \sigma^{ik} g_{kj}, \qquad \sigma_{j}^{\cdot i} = \sigma^{ki} g_{kj}, \qquad \sigma_{ij} = \sigma^{kl} g_{ki} g_{lj}.$$

Since $\sigma^{ik} = \sigma^{ki}$, there is also

$$\sigma^{i}{}_{j} = \sigma_{j}{}^{i} = \sigma^{i}_{j}, \tag{4.9}$$

but as explained on page 18, this is not the same as $\sigma^{j}_{i}$. This fact that there are nine different components $\sigma^{j}_{i}$ while there are only six different $\sigma^{ij}$ makes the components of mixed variance less desirable than the other ones.

It is easily possible to visualize the different components. We consider (in a homogeneous state of stress) an area element formed by the base vectors $\mathbf{g}_1$ and $\mathbf{g}_2$ (Figure 4.3). Because of (3.32) and (4.5) the force transmitted through it can be written as

$$d\mathbf{F} = \sigma^{ij}\, dA_i\, \mathbf{g}_j$$

with

$$d\mathbf{A} = dA_i\, \mathbf{g}^i = \mathbf{g}_1 \times \mathbf{g}_2 = \epsilon_{123}\mathbf{g}^3.$$

It follows that $dA_1 = dA_2 = 0$ and $dA_3 = \epsilon_{123}$, hence

$$d\mathbf{F} = \sigma^{3j}\epsilon_{123}\mathbf{g}_j. \tag{4.10a}$$

The stresses are, except for the factor $\epsilon_{123} = \sqrt{g}$, the contravariant components of $d\mathbf{F}$. Stresses of this kind, acting on four sides of a volume element, are shown in Figure 4.4a.

We might as well use covariant components and write

$$d\mathbf{F} = \sigma^{i}_{j}\, dA_i\, \mathbf{g}^j = \sigma^{3}_{j}\,\epsilon_{123}\,\mathbf{g}^j. \tag{4.10b}$$

Some stresses of this kind are shown in Figure 4.4b. We may also consider an

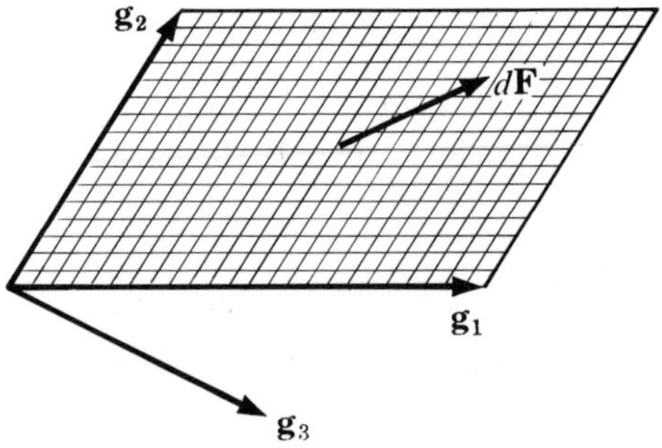


FIGURE 4.3 *Visualization of stresses.*

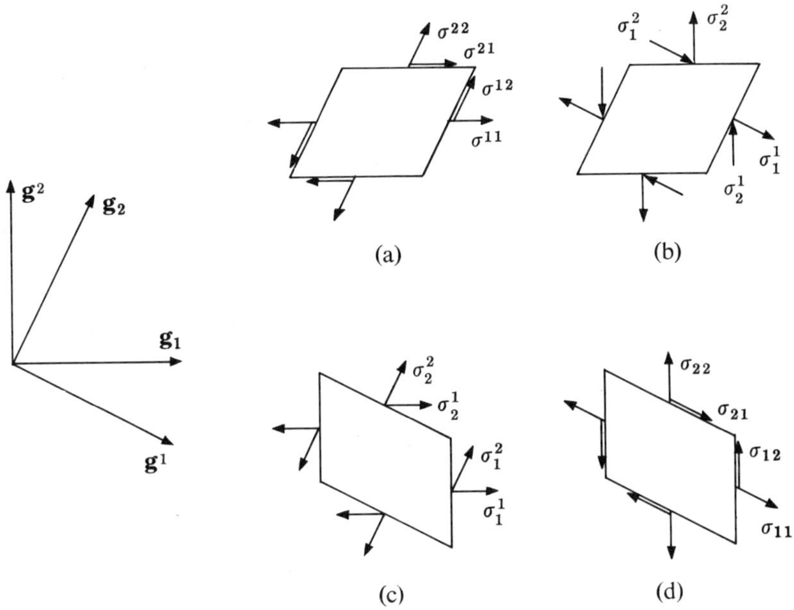


FIGURE 4.4 *Stress components of different variance.*

area element formed by the contravariant base vectors $\mathbf{g}^1$ and $\mathbf{g}^2$. In this case we use the alternative form (3.33) of the cross product and write the force transmitted through this section as

$$d\mathbf{F} = \sigma_i^j \, dA^i \, \mathbf{g}_j = \sigma_{ij} \, dA^i \, \mathbf{g}^j$$

with $d\mathbf{A} = \mathbf{g}^1 \times \mathbf{g}^2 = \epsilon^{123}\mathbf{g}_3$, whence $dA^1 = dA^2 = 0$, $dA^3 = \epsilon^{123}$ and

$$d\mathbf{F} = \sigma_3^j \epsilon^{123} \mathbf{g}_j = \sigma_{3j} \epsilon^{123} \mathbf{g}^j. \tag{4.10c, d}$$

Such stress components are illustrated by Figures 4.4c, d.

It should be noted that the stresses in Figures 4.4b, c are the same in magnitude although quite differently defined. This equality is particularly surprising for the stresses $\sigma_2^1$. On the other hand, in each of these figures it is obvious why $\sigma_2^1 \neq \sigma_1^2$, since the stresses $\sigma_1^1$ and $\sigma_2^2$ form two couples, which participate in the moment equilibrium of the element.

Among the four choices of stress components shown in Figures 4.4a–d, the first one is most appealing. The element is cut along the base vectors $\mathbf{g}_i$, which are tangent to the coordinate lines, and the forces on the sides of the element are resolved in components in the same directions. It is not surprising that these are the components of the stresses that appear in the equilibrium conditions. On the other hand, we shall see that Hooke's law assumes its simplest form when written in mixed components, since this is the compromise between contravariant stresses and the covariant strains coming out of the kinematic relations. It is one of the special merits of the tensor calculus that it affords an easy formal transition from one set of components to another one when circumstances make this desirable.

We have thus far avoided the question of the "physical components," which makes other tensor presentations of the theory of elasticity so complicated, but we may use the stresses $\sigma^{ij}$ to explain this point. As is seen from (4.10a), the force (in actual force units) on an area element $d\mathbf{A}$ is obtained by multiplying $\sigma^{ij}$ by $\epsilon_{123} = \sqrt{g}$ and by vectors $\mathbf{g}_j$, which are not unit vectors. Therefore, $\sigma^{ij}$ is not a force per unit area, but is connected with such a quantity by a conversion factor $(\sqrt{g})|\mathbf{g}_j|$, different for each component. However, having these conversion factors does not fully answer the question for the "actual" stresses, since the $\sigma^{ij}$ are skew components and what we ultimately want to have are normal and shear stresses in the usual meaning of these words. However, it would be of little use to have these orthogonal components for the nonorthogonal sections $x^i = \text{const}$ of a skew coordinate system. The most sensible thing to do is to find the principal stresses defined on page 177. They are physical components measured in stress units.

## 4.2. Constitutive Equations

The mechanics of continua draws its physical information from three sources: the conditions of equilibrium (to be replaced by Newton's law of motion in problems of dynamics), the kinematic relations, and the stress–strain relations. We have dealt in a preliminary way with the kinematic relations (p. 27) and shall come back to them after we have learned how to differentiate vectors and tensors. The equilibrium conditions will also have to be postponed until then. The stress–strain relations, however, are essentially algebraic, and after having defined stress and strain, we are ready to inspect them in detail.

The stress–strain relations are the mathematical description of the mechanical properties of the material—its constitutive equations. There are as many kinds of such equations as there are different materials, and we shall study a few of them.

We begin with a linear elastic material, but we do not require isotropy. Then every stress component $\sigma^{ij}$ is a linear function of all strain components:

$$\sigma^{ij} = E^{ijlm}\varepsilon_{lm} . \tag{4.11}$$

This is Hooke's law for a general anisotropic solid. The elastic moduli $E^{ijlm}$ represent a tensor of the fourth order. Since each of the superscripts

p72

$i, j, l, m$ can independently assume any of the values 1, 2, 3, there are $3^4 = 81$ components $E^{ijlm}$, but they are not all different from each other.

Since $\sigma^{ij} = \sigma^{ji}$, there must be $E^{ijlm} = E^{jilm}$. Since $\varepsilon_{lm} = \varepsilon_{ml}$, it is at least possible to choose $E^{ijlm} = E^{ijml}$, assigning to each of the two equal strain components half of the influence which they exert together on $\sigma^{ij}$. All together we have then

$$E^{ijlm} = E^{jilm} = E^{jiml} = E^{ijml}, \tag{4.12}$$

which says that within each pair of subscripts, $i, j$ and $l, m$, we can interchange the order. If this is permitted, there are six different pairs left (1, 1; 1, 2; 1, 3; 2, 2; 2, 3; 3, 3), and we have, at most, $6^2 = 36$ different values $E^{ijlm}$. A further reduction in number comes from the existence of a strain energy.

In cartesian coordinates we have the strain energy per unit volume (the strain energy density)

$$a = \tfrac{1}{2}(\sigma_x \varepsilon_x + \sigma_y \varepsilon_y + \sigma_z \varepsilon_z + \tau_{xy} \gamma_{xy} + \tau_{yz} \gamma_{yz} + \tau_{zx} \gamma_{zx}).$$

We bring the right-hand side of this formula in tensor form by letting $\sigma_x = \sigma^{11}, \ldots, \tau_{xy} = \sigma^{12}, \ldots, \varepsilon_x = \varepsilon_{11}, \ldots, \gamma_{xy} = \varepsilon_{12} + \varepsilon_{21}, \ldots$. This yields the expression

$$a = \tfrac{1}{2}\sigma^{ij} \varepsilon_{ij}. \tag{4.13}$$

Since this equation is in tensor form, it is valid in all coordinate systems. The left-hand side is a scalar.

Equation (4.13) shows that also in general coordinates each of the stress components does work at the corresponding strain component (if contravariant stresses and covariant strains are used), the factor $\tfrac{1}{2}$ being caused by the fact that the work on a "displacement" $\varepsilon_{ij}$ is done while a "force" is gradually increasing from zero to its final value $\sigma^{ij}$. From this we conclude that for a small increment of stress and strain the increment of $a$ is to the first order the work of the existing stress done on the increment of the strain:

$$da = \sigma^{ij}\, d\varepsilon_{ij}. \tag{4.14}$$

On the other hand, we may formally derive from (4.13) for $da$ the expression

$$da = \frac{\partial a}{\partial \sigma^{ij}}\, d\sigma^{ij} + \frac{\partial a}{\partial \varepsilon_{ij}}\, d\varepsilon_{ij}$$

and, since $\sigma^{ij}$ and $\varepsilon_{lm}$ are related through (4.11), this may be written

$$\begin{aligned} da &= \frac{\partial a}{\partial \sigma^{ij}} \frac{\partial \sigma^{ij}}{\partial \varepsilon_{lm}}\, d\varepsilon_{lm} + \frac{\partial a}{\partial \varepsilon_{ij}}\, d\varepsilon_{ij} = \frac{\partial a}{\partial \sigma^{lm}} \frac{\partial \sigma^{lm}}{\partial \varepsilon_{ij}}\, d\varepsilon_{ij} + \frac{\partial a}{\partial \varepsilon_{ij}}\, d\varepsilon_{ij} \\ &= \tfrac{1}{2}(\varepsilon_{lm} E^{lmij} + \sigma^{ij})\, d\varepsilon_{ij}. \end{aligned} \tag{4.15}$$

Because of $\varepsilon_{ij} = \varepsilon_{ji}$, the nine $d\varepsilon_{ij}$ cannot be chosen entirely arbitrarily, but we may choose one pair $ij$ and let only that $d\varepsilon_{ij} = d\varepsilon_{ji}$ be different from zero. It then follows that the sum of the coefficients of these two differentials must be the same on the right-hand sides of (4.14) and (4.15), that is,

$$\sigma^{ij} + \sigma^{ji} = \tfrac{1}{2}[\varepsilon_{lm}(E^{lmij} + E^{lmji}) + \sigma^{ij} + \sigma^{ji}].$$

Because of (4.8) and (4.12) this amounts to

$$\sigma^{ij} = \varepsilon_{lm} E^{lmij}.$$

Comparison with (4.11) shows that

$$E^{ijlm} = E^{lmij}, \tag{4.16}$$

which says that, in addition to the symmetry stated in (4.12), we are also allowed to interchange the pairs $i, j$ and $l, m$. This reduces the number of different $E^{ijlm}$ from 36 to 21.

We shall now give life to Hooke's law (4.11) and to the moduli $E^{ijlm}$ by writing explicitly the elastic laws of several materials in cartesian coordinates.

For an isotropic body we have the well-known relations

$$\sigma^{11} = \frac{E}{(1+\nu)(1-2\nu)}[(1-\nu)\varepsilon_{11} + \nu\varepsilon_{22} + \nu\varepsilon_{33}],$$

$$\sigma^{12} = \frac{E}{2(1+\nu)}(\varepsilon_{12} + \varepsilon_{21}),$$

in which we have already used tensor symbols. Comparison with (4.11) yields $E^{1111}$, $E^{1122} = E^{1133}$ and $E^{1212} = E^{1221}$ in terms of Young's modulus $E$ and Poisson's ratio $\nu$. Because of isotropy and the symmetry relations, other moduli are equal to these:

$$E^{1111} = E^{2222} = E^{3333} = \frac{E(1-\nu)}{(1+\nu)(1-2\nu)} = 2G\frac{1-\nu}{1-2\nu} = \lambda + 2\mu,$$

$$E^{1122} = E^{1133} = E^{2211}$$

$$= E^{2233} = E^{3311} = E^{3322} = \frac{E\nu}{(1+\nu)(1-2\nu)} = 2G\frac{\nu}{1-2\nu} = \lambda,$$

$$E^{1212} = E^{1221} = E^{2112} = E^{2121} = \frac{E}{2(1+\nu)} = G = \mu. \tag{4.17}$$

In the last line, eight more moduli can be added, formed with the pairs 23 and 31, but all other moduli are zero, for example

$$E^{1223} = E^{1112} = E^{1123} = 0.$$

In (4.17) the nonvanishing moduli have also been expressed in terms of the shear modulus $G$ and Poisson's ratio and in terms of Lamé's elastic constants $\lambda$ and $\mu$, which are seldom used in engineering work, but are useful in some three-dimensional problems.

One may easily verify that (4.17) can be summarized in the form

$$E^{ijlm} = \lambda\, \delta^{ij}\, \delta^{lm} + \mu(\delta^{il}\, \delta^{jm} + \delta^{im}\, \delta^{jl}). \tag{4.18}$$

Indeed, if all four superscripts are equal, all the Kronecker deltas are equal to 1, and on the right-hand side we have $\lambda + 2\mu$. If $i = j \neq l = m$, only $E^{ijlm} = \lambda$ remains, and if either $i = l \neq j = m$ or $i = m \neq j = l$, one of the $\delta$ products in the second term survives and we have $E^{ijlm} = \mu$.

Equations (4.17) and (4.18) are valid only in cartesian coordinates, but as we have done before with other equations, we may generalize (4.18) by replacing the Kronecker deltas with the proper components of the metric tensor:

$$E^{ijlm} = \lambda g^{ij} g^{lm} + \mu(g^{il} g^{jm} + g^{im} g^{jl}). \tag{4.19a}$$

Drawing upon another part of (4.17), we may write this in the alternate form

$$E^{ijlm} = \frac{E}{2(1+\nu)} \left( \frac{2\nu}{1-2\nu} g^{ij} g^{lm} + g^{il} g^{jm} + g^{im} g^{jl} \right). \tag{4.19b}$$

Both equations (4.19) are valid in any coordinate system.

As a second example, we consider an idealized porous material, as shown in Figure 4.5. It consists of many thin walls of thickness $t \ll a$, parallel

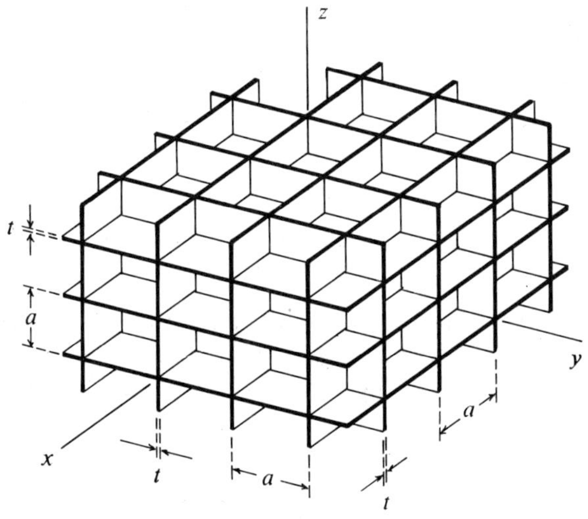


FIGURE 4.5 *Model of a porous material.*

to the coordinate planes of a cartesian system $x$, $y$, $z$. The size $a$ of the pores is small compared with the dimensions of a volume element cut from the material. In each of the walls a plane stress system is possible, and we have for example, in the walls parallel to the $x, y$ plane Hooke's law in the form

$$\left.\begin{aligned} \sigma_x &= \frac{E}{1-\nu^2}(\varepsilon_x + \nu\varepsilon_y), \\ \sigma_y &= \frac{E}{1-\nu^2}(\varepsilon_y + \nu\varepsilon_x), \\ \tau_{xy} &= G\gamma_{xy} = \frac{E}{1+\nu}\,\varepsilon_{xy}. \end{aligned}\right\} \tag{4.20}$$

To explore the elastic behavior of this porous material, we subject it to several simple states of strain and note the forces needed on the faces of a cube of side length $na$ with $n \gg 1$, Figure 4.6.

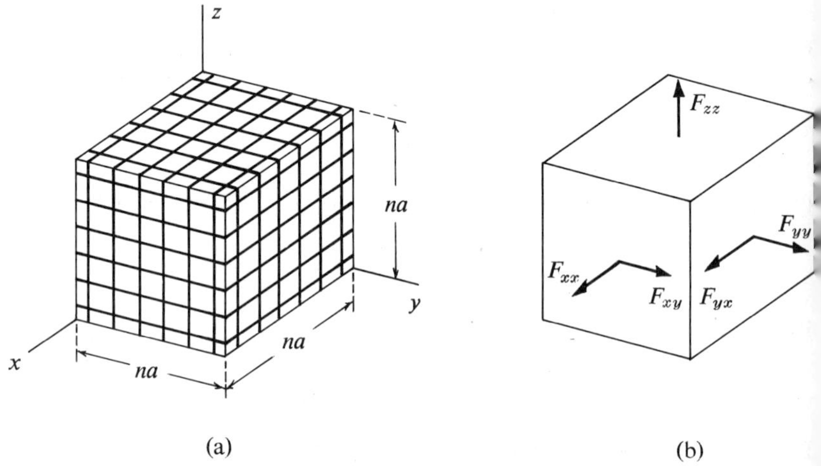


FIGURE 4.6 *Element of a porous material.*

The first strain to be imposed is a simple stretching $\varepsilon_x = \varepsilon_{11}$ in the $x$ direction with $\varepsilon_y = \varepsilon_z = 0$ and no shear strains. On the faces $x = \text{const}$ this requires stresses $\sigma_x$ in the vertical and in the horizontal walls. They add up to the force

$$F_{xx} = 2n^2 at\sigma_x = 2n^2 at \frac{E}{1-\nu^2}\,\varepsilon_{11}.$$

The vertical walls normal to the $x$ axis have no strain and no stress, and on the faces $y = \text{const}$ we have only a contribution from the horizontal walls:

$$F_{yy} = n^2 at\sigma_y = n^2 at \frac{Ev}{1 - v^2} \varepsilon_{11},$$

and a force of the same magnitude must be applied to the faces $z = \text{const}$, $F_{zz} = F_{yy}$. There are no shear forces. When we divide each force by the gross area $(na)^2$ of the face in which it is transmitted, we have the gross stresses for the porous material. We use for them the notation $\sigma^{11}$, $\sigma^{22}$, $\sigma^{33}$ and have

$$\sigma^{11} = \frac{t}{a} \frac{2E}{1 - v^2} \varepsilon_{11}, \qquad \sigma^{22} = \sigma^{33} = \frac{t}{a} \frac{Ev}{1 - v^2} \varepsilon_{11}. \tag{4.21}$$

When we subject the porous medium to a strain $\varepsilon_y = \varepsilon_{22}$ or $\varepsilon_z = \varepsilon_{33}$, similar formulas hold, and for a combined stretching in all three directions we find by linear superposition

$$\sigma^{11} = \frac{t}{a} \frac{E}{1 - v^2} (2\varepsilon_{11} + v\varepsilon_{22} + v\varepsilon_{33}) \tag{4.22a}$$

and two similar equations. Now let us subject the medium to a shear strain $\varepsilon_{xy} = \varepsilon_{12}$ in the $x$, $y$ plane. Then only the horizontal walls are stressed,

$$\tau_{xy} = \frac{E}{1 + v} \varepsilon_{12}.$$

On the faces $x = \text{const}$ and $y = \text{const}$ the stresses add up to forces

$$F_{xy} = F_{yx} = n^2 at \frac{E}{1 + v} \varepsilon_{12},$$

from which we derive the gross stress

$$\sigma^{12} = \sigma^{21} = \frac{t}{a} \frac{E}{1 + v} \varepsilon_{12} = \frac{t}{a} \frac{E}{2(1 + v)} (\varepsilon_{12} + \varepsilon_{21}). \tag{4.22b}$$

Equations (4.22a, b) and their obvious companion equations for the other coordinate planes represent the elastic law of the material in cartesian coordinates. By comparison with the general equations (4.11) we read from them the following expressions for the elastic constants:

$$\left.\begin{aligned} E^{1111} = E^{2222} = E^{3333} &= \frac{2E}{1 - v^2} \frac{t}{a}, \\ E^{1122} = E^{1133} = E^{2233} &= \frac{Ev}{1 - v^2} \frac{t}{a}, \\ E^{1212} = E^{2323} = E^{3131} &= \frac{E}{2(1 + v)} \frac{t}{a}. \end{aligned}\right\} \tag{4.23}$$

The three numerically different moduli are not independent of each other since they all depend on the two parameters $E$ and $\nu$, but their ratio is quite different from that of the corresponding moduli of an isotropic solid. With $\nu = 0.3$, (4.23) yields

$$E^{1111} : E^{1122} : E^{1212} = 1 : 0.15 : 0.175,$$

while (4.17) yields

$$E^{1111} : E^{1122} : E^{1212} = 1 : 0.428 : 0.286.$$

The porous material with cubic pores has little lateral contraction and hence needs little transverse stress to prevent it. This material is also much softer in shear than the isotropic solid.

Since the pores have the same dimension $a$ in all three directions, and since the walls have the same thickness $t$, the material shows the same elastic behavior in the directions of all three coordinate axes, but it is not isotropic. This becomes apparent when we rotate the coordinate system.

We choose a cartesian coordinate system $x^{i'}$ whose relation to a cubic cell of the material is shown in Figure 4.7. Starting from the coordinate axes $x^i$,

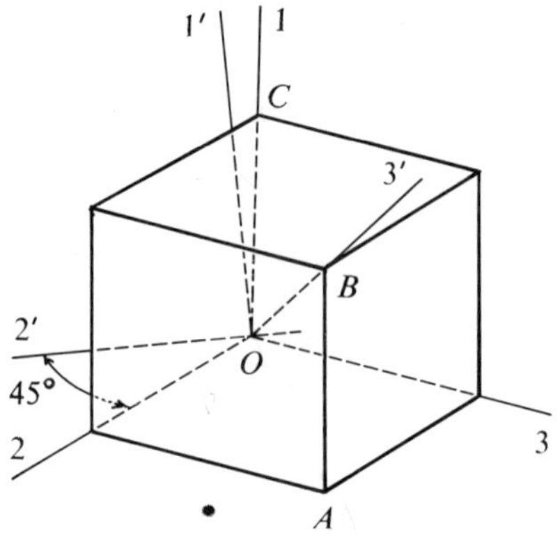


FIGURE 4.7 *Rotation of coordinate system.*

which coincide with three edges of the cube, we subject them to a rotation of 45° about the axis $x^1$. This brings the axis $x^2$ into its final position $x^{2'}$, while the axis $x^3$ assumes the position of the diagonal $OA$ of the bottom face of the cube. Then we rotate the axes about $x^{2'}$ until the axis $x^3$ coincides with a space diagonal of the cube. In this position it is the new axis $x^{3'}$ and $x^{1'}$ lies in the plane $OABC$, which is a plane of symmetry of the cellular structure. The transformation coefficients are the cosines of the angles between the axes $x^i$ and $x^{i'}$. One easily verifies the following numerical values:

$$\beta_1^{1'} = 0.816, \qquad \beta_1^{2'} = 0, \qquad \beta_1^{3'} = 0.577,$$
$$\beta_2^{1'} = -0.408, \qquad \beta_2^{2'} = 0.707, \qquad \beta_2^{3'} = 0.577,$$
$$\beta_3^{1'} = -0.408, \qquad \beta_3^{2'} = -0.707, \qquad \beta_3^{3'} = 0.577.$$

From these and the general transformation formula

$$E^{i'j'k'l'} = E^{ijkl}\beta_i^{i'}\beta_j^{j'}\beta_k^{k'}\beta_l^{l'} \tag{4.24}$$

the following values of

$$\frac{1-\nu^2}{E}\frac{a}{t}E^{i'j'k'l'}$$

may be computed:

| $k'l'$ \ $i'j'$ | 1′1′ | 2′2′ | 3′3′ | 1′2′ | 2′3′ | 3′1′ |
|---|---|---|---|---|---|---|
| 1′1′ | 1.500 | 0.467 | 0.633 | 0 | 0 | 0.236 |
| 2′2′ | | 1.500 | 0.633 | 0 | 0 | −0.236 |
| 3′3′ | | | 1.333 | 0 | 0 | 0 |
| 1′2′ | | | | 0.517 | −0.236 | 0 |
| 2′3′ | | | | | 0.683 | 0 |
| 3′1′ | | | | | | 0.683 |

The symmetry relation (4.16) may be used to fill the lower left part of the table. The numbers show that the relation between stress and strain is not the same in all directions. Also the shear moduli in the coordinate planes are not all equal. A shear strain $\varepsilon_{1'2'}$ produces not only a shear stress $\sigma^{1'2'}$, but also a shear stress $\sigma^{2'3'}$. Because of the symmetry of Figure 4.7 with respect to the 1′3′ plane, the shear stress in this plane is not coupled with shear strains in other planes. However, the strains $\varepsilon_{1'1'}$ and $\varepsilon_{2'2'}$ are coupled with the shear stress $\sigma^{3'1'} = \sigma^{1'3'}$.

The elastic law of isotropic bodies may be cast in a much simpler form, which we shall need later on. We start from (4.19b), which we introduce in (4.11):

$$\sigma^{ij} = \frac{E}{2(1+\nu)}\left(g^{il}g^{jm} + g^{im}g^{jl} + \frac{2\nu}{1-2\nu}g^{ij}g^{lm}\right)\varepsilon_{lm}.$$

We lower the index $j$ and also move the index $l$, down in the bracket and up in the strain. This yields

$$\sigma^i_j = \frac{E}{2(1+\nu)}\left(\delta^i_l\,\delta^m_j + g^{im}g_{jl} + \frac{2\nu}{1-2\nu}\,\delta^i_j\,\delta^m_l\right)\varepsilon^l_m$$

$$= \frac{E}{2(1+\nu)}\left(\varepsilon^i_j + g^{im}\varepsilon_{jm} + \frac{2\nu}{1-2\nu}\,\delta^i_j\,\varepsilon^m_m\right)$$

and finally

$$\sigma^i_j = \frac{E}{1+\nu}\left(\varepsilon^i_j + \frac{\nu}{1-2\nu}\,\varepsilon^m_m\,\delta^i_j\right). \tag{4.25}$$

A comparison with (4.17) shows that the elastic constants of this equation are the Lamé moduli:

$$\sigma^i_j = 2\mu\varepsilon^i_j + \lambda\varepsilon^m_m\,\delta^i_j\,. \tag{4.25'}$$

Equation (4.25) [just like (4.11)] gives the stress in terms of the strain. It may easily be inverted. To do so, we first let $i = j$ and invoke the summation convention for $i$. With $\delta^i_i = 3$ we have then

$$\sigma^i_i = \frac{E}{1+\nu}\left(\varepsilon^i_i + \frac{3\nu}{1-2\nu}\,\varepsilon^m_m\right) = \frac{E}{1-2\nu}\,\varepsilon^m_m\,. \tag{4.26}$$

This is used to express $\varepsilon^m_m$ in (4.25) in terms of $\sigma^m_m$:

$$\sigma^i_j - \frac{\nu}{1+\nu}\,\sigma^m_m\,\delta^i_j = \frac{E}{1+\nu}\,\varepsilon^i_j\,,$$

whence

$$E\varepsilon^i_j = (1+\nu)\sigma^i_j - \nu\sigma^m_m\,\delta^i_j\,. \tag{4.27}$$

Equations (4.25) and (4.27) represent two versions of Hooke's law in general tensor form, valid for isotropic materials. Because of the Kronecker delta, their right-hand side has only one term for shear, but two for tension. The quantities $\varepsilon^m_m$ and $\sigma^m_m$ occurring in these equations lend themselves to physical interpretation. In cartesian coordinates, $\varepsilon^m_m = \varepsilon^1_1 + \varepsilon^2_2 + \varepsilon^3_3$ is the cubic dilation, while $\frac{1}{3}\sigma^m_m$ is the average of the normal stresses and is called the hydrostatic stress. Equation (4.26) constitutes a relation between both:

$$\frac{1}{3}\,\sigma^m_m = \frac{E}{3(1-2\nu)}\,\varepsilon^m_m = K\varepsilon^m_m\,. \tag{4.28}$$

The equation states that the cubic dilatation, i.e. the elastic change of the volume, depends only on the hydrostatic stress and, in a linear material, is proportional to it. The constant $K$ is the bulk modulus of the material.

An interesting and important variant of (4.25) and (4.27) is obtained when we introduce a new notation. We write for the bulk quantities

$$3e = \varepsilon^m_m\,, \qquad 3s = \sigma^m_m \tag{4.29a, b}$$

and introduce the *strain deviation* and the *stress deviation*

$$e^i_j = \varepsilon^i_j - e\delta^i_j, \qquad s^i_j = \sigma^i_j - s\delta^i_j. \tag{4.30a, b}$$

Equation (4.26) may then be rewritten

$$Ee = (1 - 2\nu)s \tag{4.31a}$$

and (4.27) yields

$$E(e^i_j + e\delta^i_j) = (1 + \nu)s^i_j + (1 + \nu)s\delta^i_j - 3\nu s\delta^i_j = (1 + \nu)s^i_j + (1 - 2\nu)s\delta^i_j,$$

from which, after subtraction of (4.31a), there follows

$$Ee^i_j = (1 + \nu)s^i_j. \tag{4.31b}$$

For the strain deviations, we find from (4.30a) that

$$e^i_i = \varepsilon^i_i - e\delta^i_i = 3e - 3e = 0, \tag{4.32}$$

which indicates that the tensor $e^i_j$ (called the *strain deviator*) describes a change of shape without a change of volume (called a *distortion*). On the other hand, when $e^i_j = 0$, there are only three equal tensile strains

$$\varepsilon^i_j = e\delta^i_j,$$

which describe a change of volume without a change of shape (called a *dilatation*). Equations (4.29) and (4.30) describe the splitting of the strain tensor in a dilatation and a distortion and the splitting of the stress tensor in two similar parts. Equations (4.31) are that formulation of Hooke's law which uses these concepts and which turns out to be particularly simple. They may be written with the Lamé moduli:

$$s = (3\lambda + 2\mu)e, \qquad s^i_j = 2\mu e^i_j. \tag{4.33a, b}$$

After having studied elastic materials in some detail, we now turn to a few other simple materials. A *viscous fluid* is characterized by the fact that its stresses do not depend on the strains, but on the time derivatives of the strains—the *strain rates*

$$\dot{\varepsilon}_x = \frac{\partial \varepsilon_x}{\partial t}, \qquad \dot{\varepsilon}^i_j = \frac{\partial \varepsilon^i_j}{\partial t}, \qquad \dot{e} = \frac{\partial e}{\partial t}, \qquad \dot{e}^i_j = \frac{\partial e^i_j}{\partial t}.$$

For an isotropic viscous fluid, we may use (4.33), simply replacing the strains by the strain rates:

$$s = (3\lambda + 2\mu)\dot{e}, \qquad s^i_j = 2\mu\dot{e}^i_j. \tag{4.34a, b}$$

The coefficients $\lambda$ and $\mu$ are now, of course, no longer Lamé's elasticity moduli, but rather viscosity coefficients. Equation (4.34a) represents volume viscosity. In most flow problems its influence is negligible and we may get rid of it by letting $\lambda \to \infty$. Then $\dot{e} = 0$, and $s$ is independent of the deformation or vice versa. The fluid represented by this modified law is incompressible. Since real fluids combine elastic compressibility with viscous distortion, we might combine (4.33a) and (4.34b) with a proper notation for the coefficients:

$$s = 3Ke, \qquad s^i_j = 2\mu\dot{e}^i_j. \tag{4.35}$$

This leads us into the field of viscoelastic materials, which combine elastic and viscous behavior in a more general form. Their constitutive equations contain time derivatives of stress and strain and may be written in a form analogous to (4.33) and (4.34):

$$\begin{aligned} \sum_k p''_k \frac{\partial^k s}{\partial t^k} &= \sum_k q''_k \frac{\partial^k e}{\partial t^k}, \\ \sum_k p'_k \frac{\partial^k s^i_j}{\partial t^k} &= \sum_k q'_k \frac{\partial^k e^i_j}{\partial t^k}. \end{aligned} \tag{4.36}$$

In these equations, $i$ and $j$ are tensor indices, while $k$ is used as always in writing derivatives of the $k$th order. Equations (4.36) contain as special cases Hooke's law (4.31), the viscous law (4.34) and the law (4.35) of the common, compressible viscous fluid, which is elastic in compression, but viscous in distortion.

## 4.3. Plasticity

In the uni-axial tension test a material is called ideally (or perfectly) plastic if stress and strain are related according to the following rules: (i) They obey Hooke's law $\sigma = E\varepsilon$ for stresses $\sigma < \sigma_Y$. (ii) At the yield stress $\sigma_Y$ the strain can increase indefinitely, without any restriction on the strain rate $d\varepsilon/dt$ except that this rate can never be negative. (iii) The stress cannot be increased beyond $\sigma_Y$. These three statements describe the constitutive law represented in the first quadrant of Figure 4.8. A similar law holds on the negative side.

In two and three dimensions the yield limit is reached when a certain

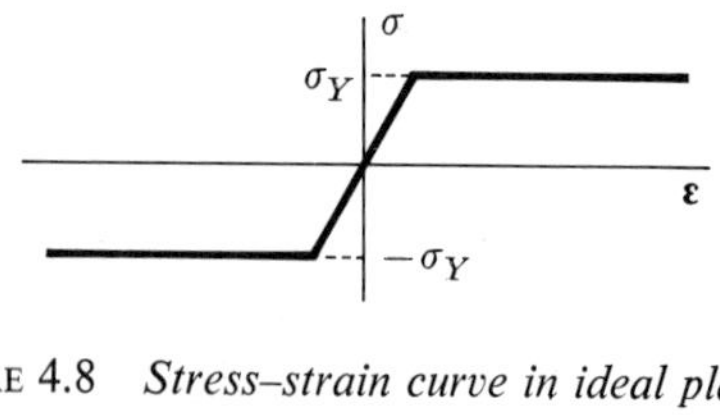


FIGURE 4.8 *Stress–strain curve in ideal plasticity.*

function of all the stress components reaches a fixed value. In the early times of the theory of plasticity attempts have been made to find "the" yield condition. Several have been proposed and experiments have been made to decide which of them is the right one. Today it is understood that there may be as many yield conditions as there are materials, and interest has been focused on establishing one which is suitable for mathematical handling and still close enough to the facts. Two such conditions are in general use, those of H. Tresca and of R. v. Mises. Between them the Mises condition is better fitted for a tensorial presentation.

No yield condition is a general law of nature, but rather a part of the constitutive equations of a particular material. Therefore, it cannot be derived by mathematical reasoning from basic principles, but it can be made plausible by correlating it with some basic concepts of mechanics.

We start from the expression (4.13) for the elastic strain energy density and there introduce the stress and strain deviators from (4.30), which we rewrite in the form

$$\sigma^{ij} = sg^{ij} + s^{ij}, \qquad \varepsilon_{ij} = eg_{ij} + e_{ij}.$$

This yields

$$a = \tfrac{1}{2}(sg^{ij} + s^{ij})(eg_{ij} + e_{ij}) = \tfrac{1}{2}(se\delta_i^i + se_i^i + s_i^i e + s^{ij}e_{ij}).$$

From (4.32), which applies equally to the stresses, it follows that the second and third terms in the parentheses vanish, and we are left with

$$a = \tfrac{1}{2}(3se + s^{ij}e_{ij}). \tag{4.37}$$

Using Hooke's law in the form (4.33), we may write this also as

$$a = \tfrac{3}{2}(3\lambda + 2\mu)ee + \mu e^{ij}e_{ij}. \tag{4.38}$$

The first term depends entirely on the change of volume and the second one on the change of shape. They are known as the *dilatation energy* and the *distortion energy*, respectively.

In plastic flow there is practically no permanent change of volume and it seems plausible that the occurrence of plastic yield should depend entirely on the stress deviator, which represents the distortional part of the stress system. The Mises yield condition in particular states that the yield limit is reached when the distortion energy

$$\frac{1}{2}\, s^{ij}e_{ij} = \frac{1}{4\mu}\, s^{ij}s_{ij} = \frac{1}{4\mu}\, s_j^i s_i^j$$

has reached a certain value, which is characteristic for the material. We may write the condition in the form

$$s_j^i s_i^j = 2k^2 \tag{4.39}$$

and we may use (4.29) and (4.30) to express the stress deviations in terms of the actual stresses and write

$$F(\sigma^i_j) \equiv 3\sigma^i_j\sigma^j_i - \sigma^i_i\sigma^j_j - 6k^2 = 0. \tag{4.40}$$

For later use we write the equation in contravariant stress components:

$$F(\sigma^{ij}) \equiv \sigma^{ik}\sigma^{jl}(3g_{jk}g_{il} - g_{ik}g_{jl}) - 6k^2 = 0. \tag{4.41}$$

In these equations, which represent the Mises yield condition, $k^2$ is the square of a material constant $k$, which has the dimension of a stress. To understand its physical meaning, we use cartesian coordinates and consider two very simple stress states.

In simple tension we have $\sigma^1_1 = \sigma_x$ and (4.40) reads

$$3\sigma_x{}^2 - \sigma_x{}^2 = 2\sigma_x{}^2 = 6k^2,$$

whence

$$\sigma_x = k\sqrt{3}.$$

In simple shear we have $\sigma^1_2 = \sigma^2_1 = \tau_{xy}$ and hence

$$3(\sigma^1_2\sigma^2_1 + \sigma^2_1\sigma^1_2) - 0 = 6\tau_{xy}{}^2 = 6k^2,$$

whence

$$\tau_{xy} = k.$$

We see that $k$ is the yield stress in simple shear and that the Mises yield condition fixes the ratio between the yield limits in shear and in uni-axial tension as $1 : \sqrt{3}$.

Once the yield limit has been reached plastic flow may begin. It is unbounded and the *flow law* regulates neither its magnitude nor its speed. As long as the stress tensor is not changed, the flow may (but need not) continue and its progress in any time element $dt$ is described by the components $\dot{\varepsilon}_{ij}\,dt$, where the dot—as elsewhere in this book—indicates differentiation with respect to time. Since we decided to neglect as insignificant the dilatational part of the plastic strain, the strain rate tensor $\dot{\varepsilon}_{ij}$ is a deviator, $\dot{\varepsilon}_{ij} = \dot{e}_{ij}$, and the work done by the stresses during the incremental strain $\dot{\varepsilon}_{ij}\,dt$ is

$$da = \dot{a}\,dt = s^{ij}\dot{\varepsilon}_{ij}\,dt = \sigma^{ij}\dot{\varepsilon}_{ij}\,dt. \tag{4.42}$$

To arrive at a reasonable flow law, we consider a change $\delta\sigma^{ij}$ of the stress tensor such that also the stresses $\sigma^{ij} + \delta\sigma^{ij}$ satisfy the yield condition. This will cause an elastic change of strain and the attendant change of strain energy, with which we are not concerned, and the subsequent plastic flow will have the strain rate $\dot{\varepsilon}_{ij} + \delta\dot{\varepsilon}_{ij}$, leading to a plastic work rate

$$\dot{a} + \delta\dot{a} = (\sigma^{ij} + \delta\sigma^{ij})(\dot{\varepsilon}_{ij} + \delta\dot{\varepsilon}_{ij}) = \dot{a} + \delta\sigma^{ij}\dot{\varepsilon}_{ij} + \sigma^{ij}\,\delta\dot{\varepsilon}_{ij}.$$

We postulate that the change $\delta\dot{a}$ depends only on the change of the deformation and not directly on the change of stress, so that

$$\delta\sigma^{ij}\dot{\varepsilon}_{ij} = 0. \tag{4.43}$$

On the other hand, we have, in view of (4.41),

$$F(\sigma^{ij} + \delta\sigma^{ij}) = F(\sigma^{ij}) + \frac{\partial F}{\partial\sigma^{ij}}\,\delta\sigma^{ij} = \frac{\partial F}{\partial\sigma^{ij}}\,\delta\sigma^{ij} = 0. \tag{4.44}$$

If every choice of the $\delta\sigma^{ij}$ which satisfies (4.44) is also to satisfy (4.43), the coefficients of both equations must be proportional to each other,

$$\dot{\varepsilon}_{ij} = \mu\,\frac{\partial F}{\partial\sigma^{ij}}, \tag{4.45}$$

where $\mu$ is an arbitrary scalar factor.

Because of the symmetry of the stress tensor the choice of the $\delta\sigma^{ij}$ is limited not only by (4.44), and we cannot say that the coefficients of $\delta\sigma^{12}$ and $\delta\sigma^{21}$ must, separately, satisfy (4.45). We may state only that

$$\dot{\varepsilon}_{12} + \dot{\varepsilon}_{21} = \mu\left(\frac{\partial F}{\partial\sigma^{12}} + \frac{\partial F}{\partial\sigma^{21}}\right),$$

but since on either side the two terms are equal to each other, this confirms the validity of (4.45).

Differentiating $F$ from (4.41) yields

$$\begin{aligned}\frac{\partial F}{\partial\sigma^{mn}} &= \sigma^{jl}(3g_{jn}g_{ml} - g_{mn}g_{jl}) + \sigma^{ik}(3g_{mk}g_{in} - g_{ik}g_{mn})\\ &= 3\sigma_{nm} - \sigma^{j}_{j}g_{mn} + 3\sigma_{nm} - \sigma^{i}_{i}g_{mn}\\ &= 6\sigma_{mn} - 6sg_{mn} = 6s_{mn}\end{aligned}$$

and (4.45) can now be written as

$$\dot{\varepsilon}_{ij} = \lambda s_{ij} \tag{4.46}$$

with $\lambda = 6\mu$. This equation is known as the Mises–Reuss flow law.

There exists an interesting way of "visualizing" the yield condition (4.41) and the flow law (4.45). We begin with introducing a new notation for the six independent stress components, letting

$$\begin{aligned}&\sigma_1 = \sigma^{11}, &&\sigma_2 = \sigma^{12} = \sigma^{21}, &&\sigma_3 = \sigma^{13} = \sigma^{31},\\ &\sigma_4 = \sigma^{22}, &&\sigma_5 = \sigma^{23} = \sigma^{32}, &&\sigma_6 = \sigma^{33}.\end{aligned}$$

We may then consider a six-dimensional *stress space*, in which $\sigma_N$ with

$N = 1, \ldots, 6$ are the cartesian coordinates of a point or the cartesian components of its position vector, the stress vector

$$\boldsymbol{\sigma} = \sigma_N \mathbf{i}_N .$$

Here $\mathbf{i}_N$ are six unit vectors serving as base vectors in the stress space and the summation convention is applied to a repeated subscript $N$ (different from the general rule, which requires a subscript and a superscript).

In this stress space the yield condition (4.41) describes a five-dimensional hypersurface, the yield surface. All points $F < 0$ lie in the interior of this surface and represent elastic states of stress. When the stresses have been increased so far that the point $\sigma_N$ comes to lie *on* the yield surface, plastic flow can occur. Points $F > 0$ represent impossible states of stress.

In the flow law (4.45) it was understood that $F$ is a function of nine stresses, $\sigma^{12}$ and $\sigma^{21}$ counting as separate variables. When we now replace $F(\sigma^{ij})$ by $F(\sigma_N)$, we have

$$\frac{\partial F}{\partial \sigma_2} = \frac{\partial F}{\partial \sigma^{12}} + \frac{\partial F}{\partial \sigma^{21}} = 2\,\frac{\partial F}{\partial \sigma^{12}}$$

and this suggests introducing a strain rate vector $\dot{\varepsilon}_N$ with the components

$$\dot{\varepsilon}_1 = \dot{\varepsilon}_{11}, \qquad \dot{\varepsilon}_2 = \dot{\varepsilon}_{12} + \dot{\varepsilon}_{21}, \qquad \dot{\varepsilon}_3 = \dot{\varepsilon}_{13} + \dot{\varepsilon}_{31},$$
$$\dot{\varepsilon}_4 = \dot{\varepsilon}_{22}, \qquad \dot{\varepsilon}_5 = \dot{\varepsilon}_{23} + \dot{\varepsilon}_{32}, \qquad \dot{\varepsilon}_6 = \dot{\varepsilon}_{33} .$$

We may then write (4.45) in the form

$$\dot{\varepsilon}_N = \mu \frac{\partial F}{\partial \sigma_N} . \tag{4.47}$$

On the left-hand side there stands a component of the strain rate vector

$$\dot{\boldsymbol{\varepsilon}} = \dot{\varepsilon}_N \mathbf{i}_N$$

and on the right-hand side the corresponding component of the vector†

$$\operatorname{grad} F = \frac{\partial F}{\partial \sigma_N} \mathbf{i}_N ,$$

which is normal to the yield surface $F = 0$. In terms of these vectors the flow law (4.47) states that the strain rate vector is normal to the yield surface.

It may be noted that, in cartesian coordinates, the components of the six-dimensional strain rate vector $\dot{\boldsymbol{\varepsilon}}$ are the usual strain components $\varepsilon_x, \ldots,$ $\gamma_{xy}, \ldots,$ and that the plastic work done during the strain increment $\dot{\boldsymbol{\varepsilon}}\, dt$ is

$$da = \sigma_N \dot{\varepsilon}_N \, dt = \boldsymbol{\sigma} \cdot \dot{\boldsymbol{\varepsilon}} \, dt . \tag{4.48}$$

† See page 70, equation (5.20).

*Problem*

4.1. The truss work shown in Figure 4.9 extends indefinitely in all directions. It consists of simple tension-compression bars connected by frictionless hinges and is used as a model for an anisotropic plane homogeneous slab. Subject this truss consecutively to uniform strains $\varepsilon_x$, $\varepsilon_y$, and $\varepsilon_{xy}$ and calculate the forces which these strains produce in the bars. Assume that all horizontal and vertical bars have the same cross section $A$ and that the diagonals have the cross section $A/2$.

Interpret the internal forces as stresses in a slab of unit thickness and calculate the elastic moduli. The relation between strains and stresses will have the form (7.27). Transform the moduli to the coordinate system $x^{\alpha'}$ shown in the figure.

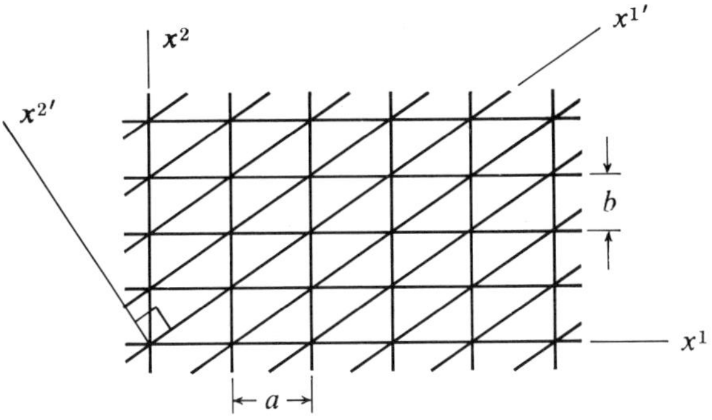


FIGURE 4.9

## References

The concept of stress and the constitutive equations of an elastic material can be found in all books on the theory of elasticity. Component presentation (cartesian, polar, etc.) is found in Love [19] and Timoshenko–Goodier [33]. Cartesian tensors are used by Eringen [6], Prager [26], and Sokolnikoff [29]; general tensors by Brillouin [3] and Green–Zerna [13]. A very thorough analysis of stress and strain of various materials is given by Prager [26] and Jaunzemis [15].

CHAPTER 5

# Derivatives and Integrals

WE HAVE DIFFERENTIATED VECTOR COMPONENTS on page 25 and have introduced the comma notation for the derivatives in definition (2.12). In rectilinear coordinates the changes of the vector components indicate the change of the vector. This is no longer so when the coordinates are curvilinear. As a simple example consider Figure 5.1a. It shows a polar coordinate system and in it three vectors **v**. The vectors are equal, but their radial components $v^1$ are not and neither are the circumferential components $v^2$. On the other hand, the vectors in Figure 5.1b are different, but they have the same radial component $v^1$ and the same circumferential component $v^2 = 0$.

## 5.1. Christoffel Symbols

When we want to differentiate a vector, we must differentiate the sum of the products of the components with the base vectors:

$$\mathbf{v}_{,j} = (v^i \mathbf{g}_i)_{,j} = v^i_{\ ,j}\, \mathbf{g}_i + v^i \mathbf{g}_{i,j} \tag{5.1a}$$

$$= (v_i \mathbf{g}^i)_{,j} = v_{i,j}\, \mathbf{g}^i + v_i\, \mathbf{g}^i_{\ ,j}\,. \tag{5.1b}$$

These formulas contain the partial derivatives of the base vectors with respect to a coordinate $x^j$. These derivatives are also vectors and may be resolved in components with respect to the base vectors $\mathbf{g}^i$ or $\mathbf{g}_i$. We write

$$\mathbf{g}_{i,j} = \Gamma_{ijk}\, \mathbf{g}^k = \Gamma^k_{ij}\, \mathbf{g}_k\,. \tag{5.2}$$

Dot-multiplying this equation by another base vector, we find

$$\mathbf{g}_{i,j} \cdot \mathbf{g}_k = \Gamma_{ijl}\, \mathbf{g}^l \cdot \mathbf{g}_k = \Gamma_{ijl}\, \delta^l_k = \Gamma_{ijk}\,, \tag{5.3a}$$

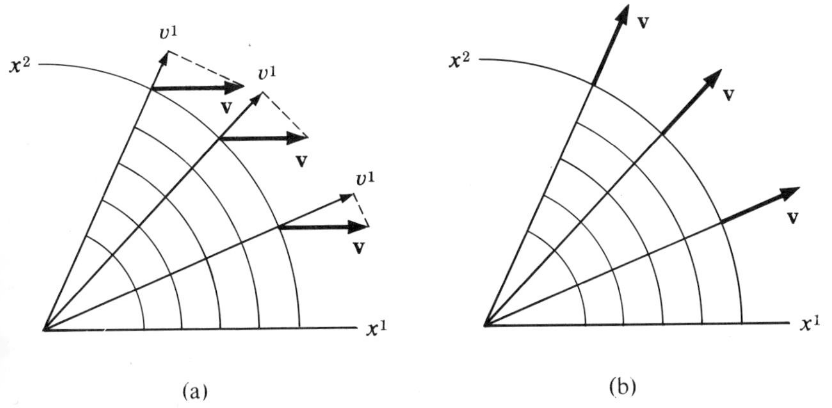


FIGURE 5.1 *Vectors in a polar coordinate system.*

$$\mathbf{g}_{i,j} \cdot \mathbf{g}^k = \Gamma^l_{ij} \mathbf{g}_l \cdot \mathbf{g}^k = \Gamma^l_{ij} \delta^k_l = \Gamma^k_{ij}. \tag{5.3b}$$

Either (5.2) or (5.3) may be considered as the definition† of the *Christoffel symbols* $\Gamma_{ijk}$ and $\Gamma^k_{ij}$. Before introducing them in the differentiation formulas (5.1) we establish a few useful relations.

Dot-multiplying the second and third members of (5.2) by $\mathbf{g}_l$ or $\mathbf{g}^l$, we find

$$\Gamma_{ijk} \delta^k_l = \Gamma_{ijl} = \Gamma^k_{ij} g_{kl} \tag{5.4a}$$

and

$$\Gamma_{ijk} g^{kl} = \Gamma^k_{ij} \delta^l_k = \Gamma^l_{ij}, \tag{5.4b}$$

which proves that the third index of the Christoffel symbols can be raised and lowered like the index of a vector component. However, this is not true for the other two subscripts, $i$ and $j$ in (5.2), and the Christoffel symbols do not transform like a third-order tensor.

When we differentiate (1.18) with respect to $x^j$, we have

$$\mathbf{g}_{i,j} = \mathbf{r}_{,ij} = \mathbf{r}_{,ji} = \mathbf{g}_{j,i} \tag{5.5}$$

and from (5.2)

$$\Gamma_{ijl} \mathbf{g}^l = \Gamma_{jil} \mathbf{g}^l, \qquad \Gamma^l_{ij} \mathbf{g}_l = \Gamma^l_{ji} \mathbf{g}_l,$$

which, after dot-multiplication by $\mathbf{g}_k$ or $\mathbf{g}^k$, yields

$$\Gamma_{ijk} = \Gamma_{jik}, \qquad \Gamma^k_{ij} = \Gamma^k_{ji}. \tag{5.6}$$

† There exists a great variety of notations for the Christoffel symbols. For details see the references on page 84.

The Christoffel symbols are symmetric with respect to the first two subscripts.
When we differentiate (1.24a) with respect to $x^k$, we find

$$g_{ij,k} = \mathbf{g}_{i,k} \cdot \mathbf{g}_j + \mathbf{g}_i \cdot \mathbf{g}_{j,k}$$

and with (5.3a)

$$g_{ij,k} = \Gamma_{ikj} + \Gamma_{jki}. \tag{5.7}$$

We write the result three times, permuting the subscripts and making use of (5.6):

$$\Gamma_{kij} + \Gamma_{jki} \qquad = g_{ij,k}, \tag{a}$$

$$\Gamma_{kij} \qquad + \Gamma_{ijk} = g_{jk,i}, \tag{b}$$

$$\Gamma_{jki} + \Gamma_{ijk} = g_{ki,j}. \tag{c}$$

Subtracting (a) from the sum of (b) and (c), we obtain the relation

$$2\Gamma_{ijk} = g_{jk,i} + g_{ki,j} - g_{ij,k}, \tag{5.8}$$

which may be used for the actual calculation of the Christoffel symbols in a coordinate system when the expressions for the components of the metric tensor are known. Formulas for the Christoffel symbols in the most common coordinate systems may be found in many textbooks and handbooks. When these formulas are used, attention must be paid to the notations, as explained in the footnote on p. 67. Since in cartesian coordinates the $g_{ij}$ are constants, it follows from (5.8) that there $\Gamma_{ijk} = \Gamma^k_{ij} = 0$.
Differentiating (1.20) with respect to $x^k$, we find that

$$\mathbf{g}_{i,k} \cdot \mathbf{g}^j + \mathbf{g}_i \cdot \mathbf{g}^j{}_{,k} = 0,$$

hence

$$\mathbf{g}_i \cdot \mathbf{g}^j{}_{,k} = -\mathbf{g}_{i,k} \cdot \mathbf{g}^j = -\Gamma^j_{ik},$$

whence, after a change of indices,

$$\mathbf{g}^i{}_{,j} = -\Gamma^i_{jk}\,\mathbf{g}^k. \tag{5.9}$$

## 5.2. Covariant Derivative

We may now return to the differentiation formulas (5.1) and write

$$\mathbf{v}_{,j} = v^i{}_{,j}\,\mathbf{g}_i + v^i\Gamma^k_{ij}\,\mathbf{g}_k.$$

By an interchanging of the notation for the two dummy indices in the last term it becomes possible to factor $\mathbf{g}_i$:

$$\mathbf{v}_{,j} = (v^i{}_{,j} + v^k\Gamma^i_{jk})\mathbf{g}_i = v^i|_j\,\mathbf{g}_i. \tag{5.10}$$

The quantity

$$v^i|_j = v^i{}_{,j} + v^k \Gamma^i_{jk} \tag{5.11}$$

is the $i$-component of the $j$-derivative of $\mathbf{v}$. As (5.11) shows, it is different from the $j$-derivative of the $i$-component, the difference being the term $v^k\Gamma^i_{jk}$. It is called the *covariant derivative* of the vector $v^i$.

A similar expression can be obtained from (5.1b) by use of (5.9):

$$\mathbf{v}_{,j} = v_{i,j}\,\mathbf{g}^i - v_i\,\Gamma^i_{jk}\,\mathbf{g}^k = (v_{i,j} - v_k\,\Gamma^k_{ij})\mathbf{g}^i = v_i|_j\,\mathbf{g}^i. \tag{5.12}$$

This leads to the covariant derivative of $v_i$:

$$v_i|_j = v_{i,j} - v_k\,\Gamma^k_{ij}\,. \tag{5.13}$$

We may bring the pairs of equations (5.10), (5.11) and (5.12), (5.13) into another form, which will be useful when we formulate differential equations of physical systems. When we proceed from a point $x^j$ to an adjacent point $x^j + dx^j$, moving along a line element vector with the components $dx^j$, a vector $\mathbf{v}(x^j)$ changes by the differential $d\mathbf{v} = \mathbf{v}_{,j}\,dx^j$ and we have

$$d\mathbf{v} = dv^i\;\mathbf{g}_i = v^i|_j\;\mathbf{g}_i\,dx^j \tag{5.14a}$$

or

$$d\mathbf{v} = dv_i\;\mathbf{g}^i = v_i|_j\,\mathbf{g}^i\,dx^j. \tag{5.14b}$$

In cartesian coordinates there is no difference between covariant and ordinary differentiation because all the $\Gamma_{ijk}$ and $\Gamma^k_{ij}$ are zero.

When we call $v^i|_j$ and $v_i|_j$ covariant derivatives, we must prove that they deserve this name. We write $\mathbf{v}$ in components in a second coordinate system $x^{i'}$:

$$\mathbf{v} = v_{i'}\,\mathbf{g}^{i'}.$$

Differentiation yields

$$\mathbf{v}_{,j'} = v_{i',j'}\,\mathbf{g}^{i'} + v_{i'}\,\mathbf{g}^{i'}{}_{,j'} = v_{i'}|_{j'}\,\mathbf{g}^{i'}.$$

On the other hand, using the chain rule of differentiation and (1.36b), we have

$$\mathbf{v}_{,j'} = \mathbf{v}_{,j}\,\frac{\partial x^j}{\partial x^{j'}} = v_i|_j\,\mathbf{g}^i\beta^j_{j'}\,.$$

Equating the right-hand sides of both equations, we see that

$$v_{i'}|_{j'}\,\mathbf{g}^{i'} = v_i|_j\,\mathbf{g}^i\beta^j_{j'}$$

and after dot-multiplication by

$$\mathbf{g}_{k'} = \beta^k_{k'}\,\mathbf{g}_k$$

it follows that

$$v_{k'}|_{j'} = v_k|_j \, \beta^k_{k'} \, \beta^j_{j'}. \tag{5.15}$$

This shows that the $v_k|_j$ are the covariant components of a second-order tensor. Similarly one proves that the $v^k|_j$ are tensor components of mixed variance. From this it follows immediately that

$$v_i|_j \, g^{ik} = v^k|_j, \tag{5.16}$$

and we may introduce the notation

$$v_i|^k = v_i|_j \, g^{jk}. \tag{5.17}$$

After having learned how to differentiate a vector, we have a brief look at the derivatives of a scalar $\phi$. We write

$$\frac{\partial \phi}{\partial x^i} = \phi_{,i}$$

and have in a second coordinate system

$$\phi_{,i'} = \frac{\partial \phi}{\partial x^{i'}} = \frac{\partial \phi}{\partial x^i} \frac{\partial x^i}{\partial x^{i'}} = \phi_{,i} \, \beta^i_{i'}, \tag{5.18}$$

where again (1.36b) has been used. Equation (5.18) shows that the $\phi_{,i}$ transform like the components of a vector. We acknowledge this by writing

$$\phi_{,i} = \phi|_i \tag{5.19}$$

and say that, for a scalar, common and covariant differentiation are identical. Since the $\phi|_i$ are vector components, they define a vector

$$\mathbf{u} = \phi|_i \, \mathbf{g}^i = \text{grad } \phi, \tag{5.20}$$

the gradient vector of the scalar field $\phi = \phi(x^i)$. The vector field $\mathbf{u} = \mathbf{u}(x^i)$ is called a gradient field. When we proceed in a scalar field $\phi(x^i)$ from the point $x^i$ to the point $x^i + dx^i$ or, in other words, when we move along the vector $dx^i$, the value of $\phi$ increases by

$$d\phi = \phi_{,i} \, dx^i = \phi|_i \, dx^i = (\text{grad } \phi) \cdot d\mathbf{x}. \tag{5.21}$$

The gradient vector measures the rate of change of a scalar field quantity and points in the direction of maximum change.

We may now turn our attention to second-order tensors. When we multiply a tensor $A_{ij}$ by two vectors $u^i$ and $v^j$, we obtain a scalar:

$$\phi = A_{ij} u^i v^j.$$

We differentiate it with respect to $x^k$ and make use of (5.11):

$$\begin{aligned}\phi_{,k} &= A_{ij,k}\, u^i v^j + A_{ij}\, u^i{}_{,k}\, v^j + A_{ij}\, u^i v^j{}_{,k} \\ &= A_{ij,k}\, u^i v^j + A_{ij}\, u^i|_k\, v^j + A_{ij}\, u^i v^j|_k \\ &\qquad - A_{ij}\, u^l \Gamma^i_{kl}\, v^j - A_{ij}\, u^i v^l \Gamma^j_{kl}.\end{aligned}$$

We can write this in the form

$$\phi|_k = (A_{ij}\, u^i v^j)|_k = A_{ij}|_k\, u^i v^j + A_{ij}\, u^i|_k\, v^j + A_{ij}\, u^i v^j|_k \tag{5.22}$$

if we define

$$A_{ij}|_k\, u^i v^j = A_{ij,k}\, u^i v^j - A_{ij}\, u^l v^j \Gamma^i_{kl} - A_{ij}\, u^i v^l \Gamma^j_{kl}.$$

By switching notations for the dummies this may be written

$$A_{ij}|_k\, u^i v^j = (A_{ij,k} - A_{lj}\, \Gamma^l_{ik} - A_{il}\, \Gamma^l_{kj}) u^i v^j.$$

This relation will hold for a certain $A_{ij}$ and for all vectors $u^i$ and $v^j$ if and only if

$$A_{ij}|_k = A_{ij,k} - A_{lj}\, \Gamma^l_{ik} - A_{il}\, \Gamma^l_{kj}, \tag{5.23a}$$

and this is the definition of the covariant derivative of $A_{ij}$.

It is important that the covariant derivative $A_{ij}|_k$ has been defined without assuming symmetry of the tensor and that its definition holds also when $A_{ij} \neq A_{ji}$.

In a similar way one may derive the following relations:

$$A^i_{\cdot j}|_k = A^i_{\cdot j,k} + A^l_{\cdot j}\, \Gamma^i_{kl} - A^i_{\cdot l}\, \Gamma^l_{jk}, \tag{5.23b}$$

$$A_i^{\cdot j}|_k = A_i^{\cdot j}{}_{,k} - A_l^{\cdot j} \Gamma^l_{ik} + A_i^{\cdot l} \Gamma^j_{kl}, \tag{5.23c}$$

$$A^{ij}|_k = A^{ij}{}_{,k} + A^{lj} \Gamma^i_{kl} + A^{il} \Gamma^j_{kl}, \tag{5.23d}$$

and similar relations hold for tensors of higher order.

The derivation of (5.23) is rather abstract and we shall now try to visualize what the covariant derivative of a tensor represents. As an example, we consider the stress tensor $\sigma^{ij}$. According to (4.5) in an arbitrary section element $d\mathbf{A} = dA_i\, \mathbf{g}^i$ located at a point $x^k$ the force

$$d\mathbf{F} = \sigma^{ij}\, dA_i\, \mathbf{g}_j$$

is transmitted. When we shift our attention to another such section element, located at an adjacent point $x^k + dx^k$, we find there a force which differs from $d\mathbf{F}$ by $d\mathbf{F}_{,k}\, dx^k$. Using (5.10) and (5.22), we write the derivative of the force vector as

$$d\mathbf{F}_{,k} = (\sigma^{ij}\, dA_i)|_k\, \mathbf{g}_j = (\sigma^{ij}|_k\, dA_i + \sigma^{ij}\, dA_i|_k)\mathbf{g}_j.$$

Now let us choose at $x^k$ and $x^k + dx^k$ two area elements $dA$ equal in size and orientation. Then $d\mathbf{A}_{,k} = 0$ and hence $dA_i|_k = 0$ and the derivative of the force $d\mathbf{F}$ is

$$d\mathbf{F}_{,k} = \sigma^{ij}|_k \, dA_i \, \mathbf{g}_j .$$

This vanishes, for any choice of the components $dA_i$ of the two section elements, if and only if $\sigma^{ij}|_k = 0$. Each of the 27 quantities $\sigma^{ij}|_k$ expresses one aspect of the rate of change of the stress tensor, namely the $j$-component of the change of the force as far as it depends on the $i$-component of $d\mathbf{A}$. Vanishing of all components of the covariant derivative means that between adjacent points the physical stress tensor (as represented by its cartesian components) does not change. A similar statement can be made for all other tensors; an example is the tensor of the elastic moduli $E^{ijlm}$ (see p. 50). If $E^{ijlm}|_k = 0$, the material is homogeneous, i.e. it has everywhere the same elastic properties with the same orientation of its anisotropy.

We arrive at two important statements when we apply covariant differentiation to the metric tensor and the permutation tensor. In cartesian coordinates $g_{ij} = \delta_{ij}$ and $\epsilon_{ijk} = e_{ijk}$ and since these are constants, we have

$$g_{ij,k} = 0 \qquad \text{and} \qquad \epsilon_{ijk,l} = 0.$$

Partial and covariant derivatives being identical in cartesian coordinates, we may just as well write

$$g_{ij}|_k = 0, \tag{5.24}$$

$$\epsilon_{ijk}|_l = 0. \tag{5.25}$$

These are two tensor equations, stating that all components of a certain tensor of the third or the fourth order are zero and, therefore, these equations carry over into all other coordinate systems. The tensors $g_{ij}$ and $\epsilon_{ijk}$ (but not their individual components!) are constants. The same can be shown to be true for $g^{ij}$ and $\epsilon^{ijk}$. Whenever one of these quantities is a factor in a product subjected to covariant differentiation, it may be pulled out of the differential operator, for example

$$(v^i g_{ij})|_k = v^i|_k g_{ij} .$$

When we apply to $\epsilon_{123}|_l$ an equation similar to (5.23a), but written for a third-order tensor, we have

$$\begin{aligned} \epsilon_{123}|_l &= \epsilon_{123,l} - \epsilon_{n23}\Gamma^n_{1l} - \epsilon_{1n3}\Gamma^n_{2l} - \epsilon_{12n}\Gamma^n_{3l} \\ &= \epsilon_{123,l} - \epsilon_{123}\Gamma^1_{1l} - \epsilon_{123}\Gamma^2_{2l} - \epsilon_{123}\Gamma^3_{3l} \\ &= \epsilon_{123,l} - \epsilon_{123}\Gamma^m_{ml} . \end{aligned}$$

Because of (5.25), this vanishes, whence

$$\epsilon_{123,l} = \epsilon_{123}\Gamma^m_{ml} . \tag{5.26}$$

Since $v_i|_j$ is a second-order covariant tensor, we may apply (5.23a) to it and form its covariant derivative with respect to $x^k$. Let

$$v_i|_j = v_{i,j} - v_m \Gamma^m_{ij} = A_{ij},$$

then

$$\begin{aligned} v_i|_j|_k &= v_i|_{jk} = A_{ij}|_k \\ &= (v_{i,j} - v_m \Gamma^m_{ij})_{,k} - (v_{l,j} - v_m \Gamma^m_{lj})\Gamma^l_{ik} - (v_{i,l} - v_m \Gamma^m_{il})\Gamma^l_{kj}. \end{aligned} \quad (5.27)$$

We ask whether this is the same as $v_i|_{kj}$, that is, whether the order of two covariant differentiations can be interchanged. We find $v_i|_{kj}$ by simply interchanging the subscripts $j$ and $k$:

$$v_i|_{kj} = (v_{i,k} - v_m \Gamma^m_{ik})_{,j} - (v_{l,k} - v_m \Gamma^m_{lk})\Gamma^l_{ij} - (v_{i,l} - v_m \Gamma^m_{il})\Gamma^l_{jk}.$$

Because of (5.6), the last term in both expressions is the same and cancels when we form the difference. The other ones yield

$$\begin{aligned} v_i|_{jk} - v_i|_{kj} = v_{i,jk} - v_{i,kj} - v_{m,k}\Gamma^m_{ij} + v_{m,j}\Gamma^m_{ik} - v_m \Gamma^m_{ij,k} \\ + v_m \Gamma^m_{ik,j} - v_{l,j}\Gamma^l_{ik} + v_{l,k}\Gamma^l_{ij} + v_m \Gamma^m_{lj}\Gamma^l_{ik} - v_m \Gamma^m_{lk}\Gamma^l_{ij}. \end{aligned}$$

On the right-hand side the first two terms cancel each other since the order of common partial differentiations is interchangeable. Among the other terms there are two pairs which cancel, and there remains

$$v_i|_{jk} - v_i|_{kj} = v_m(\Gamma^m_{ik,j} - \Gamma^m_{ij,k} + \Gamma^m_{lj}\Gamma^l_{ik} - \Gamma^m_{lk}\Gamma^l_{ij}) = v_m R^m_{\cdot ijk}. \quad (5.28)$$

Since the first member of this equation is a covariant tensor of the third order, the same must be true for the third member, and then $R^m_{\cdot ijk}$ must be a tensor of the fourth order. It is known as the *Riemann–Christoffel tensor*. Our question whether the order of two covariant differentiations is interchangeable is equivalent to the question whether $R^m_{\cdot ijk} = 0$. Since this is a tensor equation, it either holds in all coordinate systems or not at all. Now, in cartesian coordinates $\Gamma^k_{ij} = 0$ and hence also $R^m_{\cdot ijk} = 0$. This proves the interchangeability of the differentiations.

There exists, however, an important limitation to this statement. In shell theory we shall have to deal with equations which hold for the points of a curved surface, i.e. for a two-dimensional subspace of the three-dimensional space in which $R^m_{\cdot ijk} = 0$. On this curved surface a cartesian coordinate system is not possible and the argument just presented cannot be repeated. Interchangeability of covariant differentiations *on* the two-dimensional curved surface depends on whether or not a two-dimensional Riemann–Christoffel tensor $R^\mu_{\cdot\alpha\beta\gamma}$ with $\mu$, $\alpha$, $\beta$, $\gamma = 1, 2$ vanishes. This two-dimensional tensor is obtained by omitting from the definition (5.28) all terms in which one or more of the indices have the value 3. It is possible and really happens that the remainder does not add up to zero and then it matters which of two differentiations is carried out first. We shall come back to this point on page 140.

The reader may expect that there exists a relation similar to (5.12) between the second derivative $\mathbf{v}_{,jk}$ of a vector $\mathbf{v}$ and the second covariant derivative $v_i|_{jk}$. It is easy to show and important to know that the relation is more complicated. To find it, we start from the second member of (5.12) and differentiate it with respect to $x^k$:

$$\mathbf{v}_{,jk} = v_{i,jk}\,\mathbf{g}^i + v_{i,j}\,\mathbf{g}^i{}_{,k} - v_{i,k}\,\Gamma^i_{jl}\,\mathbf{g}^l - v_i\,\Gamma^i_{jl,k}\,\mathbf{g}^l - v_i\,\Gamma^i_{jl}\,\mathbf{g}^l{}_{,k}.$$

Making use of (5.9) and changing the notation of dummy indices where convenient, we bring this into the following form:

$$\mathbf{v}_{,jk} = (v_{i,jk} - v_{l,j}\,\Gamma^l_{ik} - v_{m,k}\,\Gamma^m_{ij} - v_m\,\Gamma^m_{ij,k} + v_m\,\Gamma^m_{lj}\,\Gamma^l_{ik})\mathbf{g}^i.$$

In the parentheses we recognize part of the right-hand side of (5.27) and hence we have

$$\mathbf{v}_{,jk} = v_i|_{jk}\,\mathbf{g}^i + (v_{i,l} - v_m\,\Gamma^m_{il})\Gamma^l_{kj}\,\mathbf{g}^i,$$

which, with the help of (5.13), may ultimately be written as

$$\mathbf{v}_{,jk} = (v_i|_{jk} + v_i|_l\,\Gamma^l_{jk})\mathbf{g}^i \tag{5.29a}$$

and in the alternate form

$$\mathbf{v}_{,jk} = (v^i|_{jk} + v^i|_l\,\Gamma^l_{jk})\mathbf{g}_i. \tag{5.29b}$$

## 5.3. Divergence and Curl

In (5.20) we saw the tensor form of the gradient. It is a vector whose covariant components are the covariant derivatives of a scalar function $\phi(x^i)$. We can easily find similar expressions for the other two field derivatives used in the Gibbs form of vector analysis, the divergence and the curl.

In cartesian coordinates the divergence of a vector field $\mathbf{v}$ is

$$\operatorname{div}\mathbf{v} = \frac{\partial v_x}{\partial x} + \frac{\partial v_y}{\partial y} + \frac{\partial v_z}{\partial z}.$$

To bring this formula into tensor form, we replace $\partial v_x/\partial x$ by $v^1{}_{,1} = v^1|_1$ etc. and find

$$\operatorname{div}\mathbf{v} = v^i|_i = v_i|^i \tag{5.30}$$

as an expression valid in all coordinate systems.

In cartesian coordinates the curl is defined as

$$\operatorname{curl}\mathbf{v} = \left(\frac{\partial v_z}{\partial y} - \frac{\partial v_y}{\partial z}\right)\mathbf{i} + \left(\frac{\partial v_x}{\partial z} - \frac{\partial v_z}{\partial x}\right)\mathbf{j} + \left(\frac{\partial v_y}{\partial x} - \frac{\partial v_x}{\partial y}\right)\mathbf{k}.$$

Each of the six terms of this expression involves three directions, that of a component of **v**, the direction in which the derivative is taken, and that of the unit vector with which this derivative is multiplied. These are always three different directions, and when any two of them are interchanged, a term with the opposite sign results. This points to the use of the permutation symbols, and we rewrite the curl in the following form:

$$\operatorname{curl}\mathbf{v} = \left(\frac{\partial v_z}{\partial y}\,e_{yzx} + \frac{\partial v_y}{\partial z}\,e_{zyx}\right)\mathbf{i} + \left(\frac{\partial v_x}{\partial z}\,e_{zxy} + \frac{\partial v_z}{\partial x}\,e_{xzy}\right)\mathbf{j}$$

$$+ \left(\frac{\partial v_y}{\partial x}\,e_{xyz} + \frac{\partial v_x}{\partial y}\,e_{yxz}\right)\mathbf{k}.$$

In cartesian coordinates, the permutation symbols are identical with the components of the permutation tensor, and the unit vectors **i**, **j**, **k** are the base vectors $\mathbf{g}_i$ or $\mathbf{g}^i$. We may therefore write

$$\operatorname{curl}\mathbf{v} = \left(\frac{\partial v_3}{\partial x^2}\,\epsilon^{231} + \frac{\partial v_2}{\partial x^3}\,\epsilon^{321}\right)\mathbf{g}_1 + \left(\frac{\partial v_1}{\partial x^3}\,\epsilon^{312} + \frac{\partial v_3}{\partial x^1}\,\epsilon^{132}\right)\mathbf{g}_2$$

$$+ \left(\frac{\partial v_2}{\partial x^1}\,\epsilon^{123} + \frac{\partial v_1}{\partial x^2}\,\epsilon^{213}\right)\mathbf{g}_3\,.$$

Now we replace the common partial derivatives by covariant derivatives and make use of the summation convention and thus arrive at the simple form

$$\mathbf{w} = \operatorname{curl}\mathbf{v} = v_j|_i\,\epsilon^{ijk}\mathbf{g}_k = v^j|^i\epsilon_{ijk}\,\mathbf{g}^k \tag{5.31}$$

which, having tensor form, is generally valid. In pure tensor notation it says that the vector

$$w^k = v_j|_i\,\epsilon^{ijk} \tag{5.31'}$$

is the curl of the vector $v_j$. The covariant derivative $v_j|_i$ of a vector $v_j$ is a tensor. The curl is the vector associated with the antimetric part of this tensor, introduced in (3.24).

Another differential operator which belongs in this group is the Laplace operator. It may be defined in several ways, for example as the divergence of a gradient:

$$\nabla^2\phi = \operatorname{div}(\operatorname{grad}\phi).$$

We write (5.20) in the form

$$\mathbf{u} = \operatorname{grad}\phi = \phi|_i\,\mathbf{g}^i = u_i\,\mathbf{g}^i$$

and apply (5.30) to obtain

$$\nabla^2\phi = \operatorname{div}\mathbf{u} = u_i|^i = \phi|_i^i\,. \tag{5.32}$$

Now let us take the curl of (5.20). Using (5.31), we find that

$$\text{curl grad } \phi = \phi|_j|_i \epsilon^{ijk} \mathbf{g}_k = \phi|_{ji} \epsilon^{ijk} \mathbf{g}_k .$$

Since $\phi|_{ji} = \phi|_{ij}$, this vanishes:

$$\text{curl grad } \phi = 0. \tag{5.33}$$

A gradient field has no curl.

Another equation of this kind is found by taking the divergence of (5.31):

$$\text{div curl } \mathbf{v} = (v_j|_i \epsilon^{ijk})|_k = v_j|_{ik} \epsilon^{ijk}.$$

Since $v_j|_{ik}$ is symmetric with respect to $i$ and $k$, also this vanishes:

$$\text{div curl } \mathbf{v} = 0. \tag{5.34}$$

A vector field $\mathbf{w} = \text{curl } \mathbf{v}$ is called a *solenoidal field* and $\mathbf{v}$ is called the *vector potential* of the field $\mathbf{w}$. We see that a solenoidal field has no divergence.

## 5.4. The Integral Theorems of Stokes and Gauss

Consider a plane or a curved surface $S$ and in it a coordinate system $x^\alpha$. On this surface we draw a curve $C$ and select in it a line element

$$d\mathbf{s} = ds^\alpha \ \mathbf{g}_\alpha .$$

If $C$ is a closed curve and the contour of a domain $A$, we shall always choose the direction of $d\mathbf{s}$ such that the interior of $A$ is to its left.† To every line element $d\mathbf{s}$ we define a normal vector $d\mathbf{n}$ which has the direction of the outer normal, is tangential to $S$, and has the same length as $d\mathbf{s}$. Assuming that in Figure 5.2 a unit vector $\mathbf{i}_3$ is normal to $S$ and points out of the paper, we may write

$$d\mathbf{n} = d\mathbf{s} \times \mathbf{i}_3 = ds^\alpha \ \epsilon_{\alpha 3 \beta} \mathbf{g}^\beta = dn_\beta \ \mathbf{g}^\beta,$$

whence

$$dn_\beta = ds^\alpha \epsilon_{\beta\alpha} . \tag{5.35}$$

Now let there be in $S$ a vector field $\mathbf{u} = u^\gamma \mathbf{g}_\gamma$. We want to evaluate

$$\oint \mathbf{u} \cdot d\mathbf{n} = \oint u^\gamma \ dn_\gamma$$

along the contour of the small quadrilateral $ABCD$ in Figure 5.3, following it counterclockwise as indicated by the little arrows. On each of the four sides

† This definition implies that one side of the surface is considered the upper side. In pictures it is always obvious which side this is; but, when one faces the real object in space, this is not so and an additional convention is required to give our statement meaning.

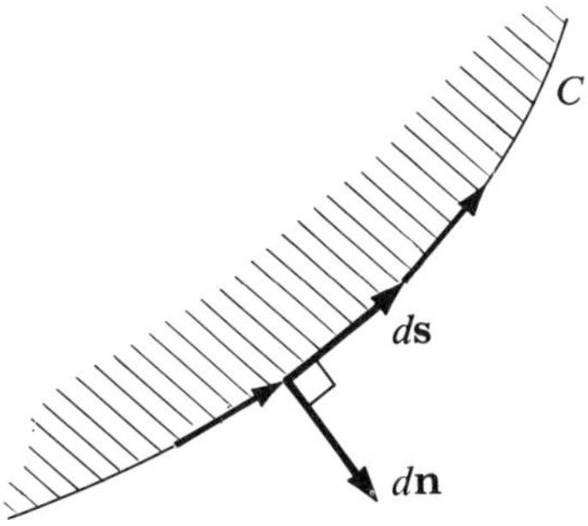


FIGURE 5.2 *Normal vector of a line element.*

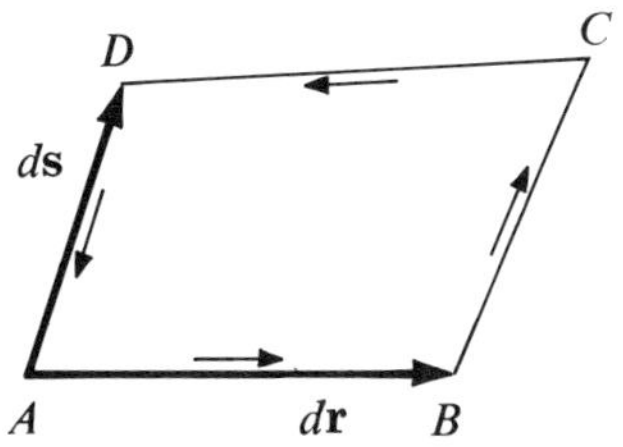


FIGURE 5.3 *Area element.*

we choose for **u** the value at one end as representative. This amounts to disregarding contributions which are in the differentials of a higher order than the terms carried. Using (5.35), we find

$$\oint u^\gamma\, dn_\gamma = u^\gamma\, dr^\alpha \epsilon_{\gamma\alpha} + (u^\gamma + u^\gamma|_\beta\, dr^\beta)\, ds^\alpha \epsilon_{\gamma\alpha} - (u^\gamma + u^\gamma|_\beta\, ds^\beta)\, dr^\alpha \epsilon_{\gamma\alpha} - u^\gamma\, ds^\alpha \epsilon_{\gamma\alpha}$$
$$= u^\gamma|_\beta\, \epsilon_{\gamma\alpha}(dr^\beta\, ds^\alpha - dr^\alpha\, ds^\beta).$$

We change dummy indices and then apply (3.30) and (3.29):

$$\oint u^\gamma\, dn_\gamma = (u^\gamma|_\alpha\, \epsilon_{\gamma\beta} - u^\gamma|_\beta\, \epsilon_{\gamma\alpha})\, dr^\alpha\, ds^\beta = u^\gamma|_\delta\, \epsilon_{\gamma\zeta}\epsilon^{\delta\zeta}\epsilon_{\alpha\beta}\, dr^\alpha\, ds^\beta$$
$$= u^\gamma|_\delta\, \delta^\delta_\gamma\, \epsilon_{\alpha\beta}\, dr^\alpha\, ds^\beta = u^\gamma|_\gamma\, \epsilon_{\alpha\beta}\, dr^\alpha\, ds^\beta. \tag{5.36}$$

In a reference frame consisting of the base vectors $\mathbf{g}^\alpha$ and $\mathbf{i}^3$ the area of the quadrilateral $ABCD$ is described by a vector $d\mathbf{A} = dA_3\, \mathbf{i}^3$, and since there are no other components, we may write $dA_3 = dA$. From (3.43) we have then

$$dA = dr^\alpha\, ds^\beta\, \epsilon_{\alpha\beta}$$

and hence

$$\oint u^\gamma\, dn_\gamma = u^\gamma|_\gamma\, dA.$$

Many such quadrilaterals may be assembled to fill the area $A$ enclosed by a curve $C$, Figure 5.4. In the sum of the contour integrals the contributions of all interior elements $d\mathbf{r}$ and $d\mathbf{s}$ cancel because they appear twice with opposite signs in adjacent area elements. Only the contributions of line elements of the contour $C$ of $A$ are left and their sum, the integral of $u^\gamma\, dn_\gamma$ around $C$, equals the sum of the contributions $u^\gamma|_\gamma\, dA$ of all area elements:

$$\int_A u^\gamma|_\gamma\, \epsilon_{\alpha\beta}\, dr^\alpha\, ds^\beta = \oint_C u^\gamma\, dn_\gamma. \tag{5.37}$$

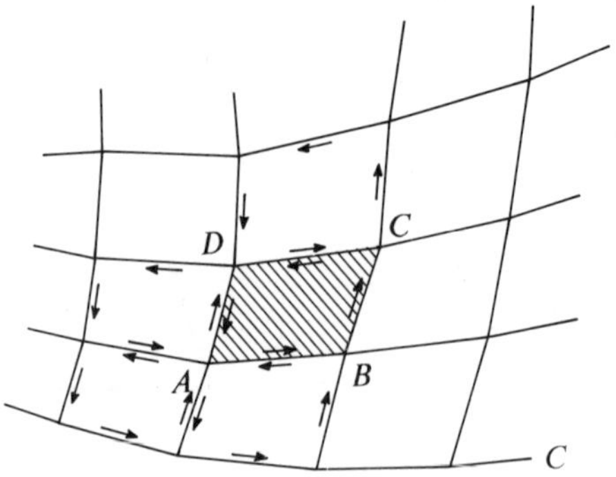


FIGURE 5.4 *Area elements and boundary curve C.*

This equation is the two-dimensional form of Gauss' *divergence theorem.* We may write it also in the Gibbs notation:

$$\int_A \operatorname{div} \mathbf{u}\, dA = \oint_C \mathbf{u} \cdot d\mathbf{n}. \tag{5.37'}$$

A similar relation can be derived for a three-dimensional vector field $\mathbf{u} = u^l \mathbf{g}_l$. We choose a volume element with edges $dr^i\, \mathbf{g}_i$, $ds^j\, \mathbf{g}_j$, and $dt^k\, \mathbf{g}_k$ as shown in Figure 5.5 and calculate for it the surface integral

$$\int \mathbf{u} \cdot d\mathbf{A} = \int u^l\, dA_l,$$

where each of the vectors $dA_l\, \mathbf{g}^l$ representing one of the six faces is assumed to have the direction of the outer normal.

For the face on the right we have $dA_l = ds^j\, dt^k\, \epsilon_{jkl}$ and with the opposite sign the same expression represents the face on the left. The corresponding values of the vector components are $u^l$ on the left and $u^l + u^l|_i\, dr^i$ on the right. The sum of the contributions of these two faces is

$$u^l(-dA_l) + (u^l + u^l|_i\, dr^i)\, dA_l = u^l|_i\, dr^i\, dA_l = u^l|_i\, dr^i\, ds^j\, dt^k\, \epsilon_{jkl}.$$

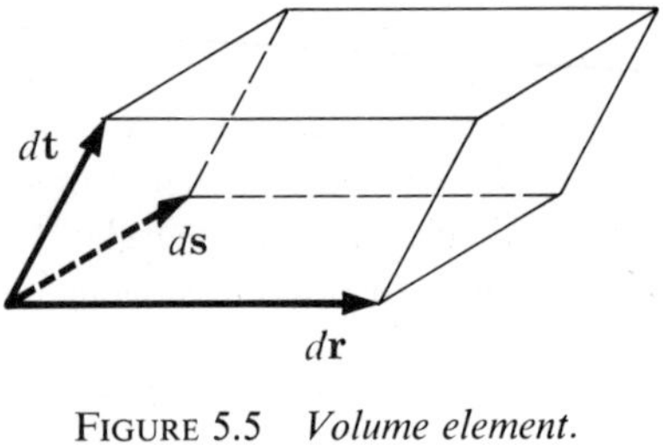


FIGURE 5.5 *Volume element.*

Applying (3.21) and then (3.19), we bring this successively into the following forms:

$$\begin{aligned} u^l|_i \epsilon_{ljk} &= u^l|_m \delta_i^m \epsilon_{ljk} = \tfrac{1}{2} u^l|_m \epsilon^{mpq} \epsilon_{ipq} \epsilon_{ljk} \\ &= \tfrac{1}{2} u^l|_m \epsilon_{ipq} (\delta_l^m \delta_j^p \delta_k^q - \delta_l^m \delta_k^p \delta_j^q + \delta_k^m \delta_l^p \delta_j^q - \delta_k^m \delta_j^p \delta_l^q + \delta_j^m \delta_k^p \delta_l^q - \delta_j^m \delta_l^p \delta_k^q) \\ &= \tfrac{1}{2} [u^l|_l (\epsilon_{ijk} - \epsilon_{ikj}) + u^l|_k (\epsilon_{ilj} - \epsilon_{ijl}) + u^l|_j (\epsilon_{ikl} - \epsilon_{ilk})] \\ &= u^l|_l \epsilon_{ijk} - u^l|_k \epsilon_{ijl} - u^l|_j \epsilon_{ilk}. \end{aligned}$$

The contributions of the other two pairs of faces of the volume element are obtained by cyclic permutation of the free indices $ijk$ and the total is

$$\begin{aligned} \int u^l \, dA_l &= (u^l|_l \epsilon_{ijk} - u^l|_k \epsilon_{ijl} - u^l|_j \epsilon_{ilk} + u^l|_l \epsilon_{jki} - u^l|_i \epsilon_{jkl} - u^l|_k \epsilon_{jli} \\ &\quad + u^l|_l \epsilon_{kij} - u^l|_j \epsilon_{kil} - u^l|_i \epsilon_{klj}) \, dr^i \, ds^j \, dt^k \\ &= (3u^l|_l \epsilon_{ijk} - 2u^l|_i \epsilon_{ljk} - 2u^l|_j \epsilon_{ilk} - 2u^l|_k \epsilon_{ijl}) \, dr^i \, ds^j \, dt^k. \end{aligned} \tag{5.38}$$

In the second, third, and fourth terms the sum over $l$ contains at most one term because $l$ must be different from the other two indices of the $\epsilon$ factor. If there is no difference between these two indices, the term drops out altogether. We consider two typical sequences $i, j, k$. When all three indices are different, for example $i, j, k = 1, 2, 3$, we have

$$(\,) = 3u^l|_l \epsilon_{123} - 2u^1|_1 \epsilon_{123} - 2u^2|_2 \epsilon_{123} - 2u^3|_3 \epsilon_{123} = u^l|_l \epsilon_{123}$$

and when two of them are equal, for example, $i, j, k = 2, 2, 3$, we have

$$(\,) = 0 - 2u^1|_2 \epsilon_{123} - 2u^1|_2 \epsilon_{213} - 0 = 0.$$

We see that the expression enclosed in parentheses in (5.38) is nonzero only if $i \neq j \neq k$ and then equals $u^l|_l \epsilon_{ijk}$. Hence,

$$\int u^l \, dA_l = u^l|_l \epsilon_{ijk} \, dr^i \, ds^j \, dt^k. \tag{5.39}$$

Comparison of the right-hand side with (3.44) and (5.30) shows that it is the product of div **u** and the volume $dV$ of the element.

When we combine volume elements to form a finite domain $V$, the contributions of faces common to two adjacent elements cancel from the sum of all integrals and we are left with the integral over the surface $S$ of the domain, and (5.39) shows that

$$\int_V u^l|_l \epsilon_{ijk} \, dr^i \, ds^j \, dt^k = \int_S u^l \, dA_l \tag{5.40}$$

or, in Gibbs notation,

$$\int_V \operatorname{div} \mathbf{u} \, dV = \int_S \mathbf{u} \cdot d\mathbf{A}. \tag{5.40'}$$

This is the three-dimensional form of the divergence theorem of Gauss.

Both versions (5.37) and (5.40) of the theorem state that in any vector field **u** the integral of its divergence, extended over a closed domain, equals the integral of the normal component, extended over the boundary.

Let us consider once more a vector field $\mathbf{u} = u_i \mathbf{g}^i$ and place in it the infinitesimal triangle shown in Figure 5.6. This triangle is represented by the area vector

$$d\mathbf{A} = \tfrac{1}{2} d\mathbf{r} \times d\mathbf{t} = \tfrac{1}{2} dr^i \, dt^j \, \epsilon_{ijk} \mathbf{g}^k. \tag{5.41}$$

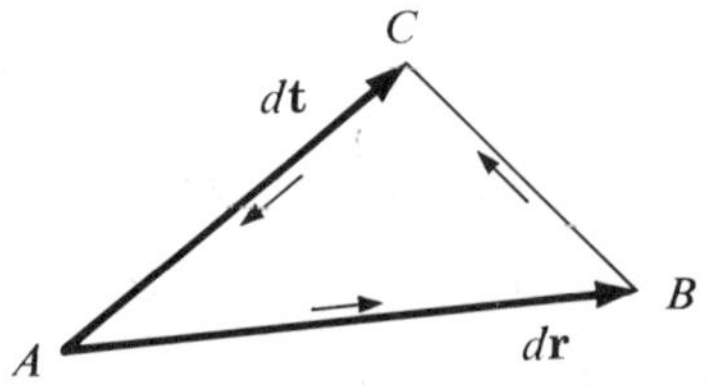


FIGURE 5.6 *Triangular area element.*

We calculate the integral $\oint u_i \, ds^i$ extended counterclockwise over the contour *ABCA*. When we follow the little arrows, $ds^i$ is in sequence identical with $dr^i$, $dt^i - dr^i$, and $-dt^i$. On each of these sides we use for $u_i$ the average of its value at the endpoints and find

$$\begin{aligned}\oint u_i \, ds^i &= (u_i + \tfrac{1}{2} u_i|_j \, dr^j) \, dr^i + [u_i + \tfrac{1}{2} u_i|_j (dr^j + dt^j)](dt^i - dr^i) \\ &\quad + (u_i + \tfrac{1}{2} u_i|_j \, dt^j)(-dt^i) \\ &= \tfrac{1}{2} u_i|_j (dr^j \, dt^i - dr^i \, dt^j) = u_i|_j (\delta^j_m \delta^i_n - \delta^j_n \delta^i_m) \tfrac{1}{2} dr^m \, dt^n.\end{aligned}$$

To the last expression we apply (3.20) and this leads to the result

$$\oint u_i \, ds^i = u_i|_j \, \epsilon^{jik} \epsilon_{mnk} \tfrac{1}{2} dr^m \, dt^n.$$

Comparison with (5.31) and (5.41) shows that this is the product of the $k$-component of curl **u** and the component $dA_k$ of the area vector of the triangle and, after summation over $k$, it is the product (curl **u**) $\cdot \, d\mathbf{A}$.

Any area $A$ with contour $C$ in a plane or on a curved surface may be covered with triangular elements like the one just considered. In the sum of all their contour integrals of $u_i \, ds^i$ the contributions of interior boundaries cancel again and only the integral over the contour $C$ is left and this equals the integral of the right-hand side extended over the entire domain $A$:

$$\oint_C u_i \, ds^i = \tfrac{1}{2} \epsilon^{jik} \epsilon_{mnk} \int_A u_i|_j \, dr^m \, dt^n \tag{5.42}$$

or, in Gibbs notation,

$$\oint_C \mathbf{u} \cdot d\mathbf{s} = \int_A (\text{curl } \mathbf{u}) \cdot d\mathbf{A}. \tag{5.42'}$$

This is *Stokes' theorem.* The integral on the left-hand side is called the *circulation* of $\mathbf{u}$ along the curve $C$.

In (5.42) the components of the permutation tensor have been pulled before the integral sign although each one of them depends on the space coordinates in the domain $A$. This was possible since the product $\epsilon^{jik}\epsilon_{mnk} = e^{jik}e_{mnk}$ and has only the values $\pm 1$ or 0 and, hence, is a constant.

On page 70, we learned how to derive from every scalar field $\phi = \phi(x^i)$ a vector field

$$\mathbf{u} = \text{grad } \phi = \phi|_i \mathbf{g}^i = \phi_{,i}\mathbf{g}^i. \tag{5.43}$$

We invert the problem and, starting from a given vector field $\mathbf{u}$, try to find a scalar field $\phi$ of which it is the gradient. The problem does not necessarily have a solution and we shall see under which condition it does.

We choose a fixed point $O$ and a variable point $P$, which may take any position in space, connect both by a curve and integrate:

$$\int_O^P \mathbf{u} \cdot d\mathbf{s} = \int_O^P u_i \, dx^i = \int_O^P \phi_{,i} \, dx^i = \phi(P) - \phi(O). \tag{5.44}$$

In order to yield a unique value of $\phi$ for every point $P$ it is necessary that the integral is independent of the path chosen for the integration. If we choose two paths, $OQP$ and $ORP$, and the integral for both of them is the same, then the integral over the closed loop $OQPRO$ vanishes. This is the condition for the existence of a potential $\phi$ of a vector field $\mathbf{u}$. It amounts to requiring that the integral on the left-hand side of (5.42) must vanish for any closed curve $C$, and this in turn is only possible if curl $\mathbf{u} = 0$ everywhere. On page 76 we saw that in every gradient field the curl vanishes. Here we see that the opposite is also true:

$$\text{If} \quad \text{curl } \mathbf{u} = 0, \qquad \text{then} \quad \mathbf{u} = \text{grad } \phi; \tag{5.45}$$

every vector whose curl vanishes is a gradient.

It is also possible to prove the inverse of (5.34), namely, that any vector field $\mathbf{w}$ whose divergence vanishes is a solenoidal field and has a vector potential $\mathbf{v}$ connected with it by (5.31) or (5.31'). It suffices to prove the theorem in cartesian coordinates. Since it is a statement about vectors, it will then hold true in all coordinate systems.

In cartesian coordinates $x^i$, (5.31') reads

$$v_{3,2} - v_{2,3} = w^1, \tag{a}$$
$$v_{1,3} - v_{3,1} = w^2, \tag{b}$$
$$v_{2,1} - v_{1,2} = w^3, \tag{c}$$

and we consider these as three simultaneous partial differential equations for the unknown components $v_i$ of the vector potential. To construct a particular solution, we arbitrarily let

$$v_3 \equiv 0. \tag{d}$$

Then, from (b) and (a),

$$v_1 = \int_0^{x^3} w^2 \, dx^3, \qquad v_2 = -\int_0^{x^3} w^1 \, dx^3 + f(x_1, x_2), \tag{e, f}$$

the integrals being extended along any line $x^1 = \text{const}$, $x^2 = \text{const}$ (that is, a coordinate line $x^3$). Upon introducing these expressions for $v_1$ and $v_2$ in (c), we find that

$$-\int_0^{x^3} (w^2{}_{,2} + w^1{}_{,1}) \, dx^3 + f_{,1} = w^3.$$

We now make use of our assumption that div $\mathbf{w} = 0$, that is, that

$$w^1{}_{,1} + w^2{}_{,2} + w^3{}_{,3} = 0$$

and find that, for fixed values of $x^1$ and $x^2$,

$$\int_0^{x^3} w^3{}_{,3} \, dx^3 + f_{,1} = w^3(x^3).$$

For $x^3 = 0$ this yields

$$f_{,1}(x^1, x^2) = w^3(x^1, x^2, 0)$$

and then

$$f(x^1, x^2) = \int_0^{x^1} w^3 \, dx^1.$$

When this is inserted in (f), this equation together with (d) and (e) represents in cartesian coordinates a vector potential $\mathbf{v}$ of the field $\mathbf{w}$. It is not the only one. In view of (5.33), we may add to it an arbitrary gradient field $\mathbf{u} = \text{grad}\,\phi$ and thus have a more general vector potential $\boldsymbol{\omega} = \mathbf{v} + \mathbf{u}$ of the field $\mathbf{w}$. This proves that every vector field whose divergence vanishes can be written as the curl of another field:

$$\text{If} \quad \text{div}\,\mathbf{w} = 0, \qquad \text{then} \quad \mathbf{w} = \text{curl}\,\boldsymbol{\omega}. \tag{5.46}$$

The result is a property of a vector field and this may be restated in any mathematical language, in particular in tensor notation, valid in all coordinate systems.

Now let us consider an arbitrary vector field $\mathbf{u} = u^i\mathbf{g}_i$ and calculate its divergence and its curl:

$$u^i|_i = \delta, \qquad u^j|^i\epsilon_{ijk} = \gamma_k. \tag{5.47}$$

We propose to find a vector field $v^i$ which has the same divergence but no curl:

$$v^i|_i = \delta, \qquad v^j|^i\epsilon_{ijk} = 0.$$

This is a gradient field

$$v_i = \phi|_i$$

and hence

$$v_i|^i = \phi|^i_i = \delta. \tag{5.48}$$

This is Poisson's equation. It always has a solution, and the solution is not even unique unless we subject it to a boundary condition. The difference

$$w^i = u^i - v^i \tag{5.49}$$

has no divergence, but the same curl as $u^i$:

$$w^i|_i = 0, \qquad w^j|^i\epsilon_{ijk} = \gamma_k$$

and is a solenoidal field. This proves that every vector field may be split (in more than one way) into a gradient field and a solenoidal field and hence may be written in the form

$$u_i = \phi|_i + \omega^k|^j\epsilon_{ijk}, \tag{5.50}$$

where $\phi$ is a scalar and $\omega_k$ a vector potential.

## *Problems*

5.1. For the vectors shown in Figure 5.1a, show that $v_i|_j = 0$ while $v_{i,j} \neq 0$.

5.2. For the vectors shown in Figure 5.1b, show that $v_i|_j \neq 0$ while $v_{i,j} = 0$.

5.3. In a plane polar coordinate system $r = x^1$, $\theta = x^2$ a vector field is described by the equations

$$v^1 = A\cos x^2, \qquad v^2 = -(A/x^1)\sin x^2.$$

Find the covariant derivatives $v^\alpha|_\beta$.

5.4. For spherical coordinates $r = x^1$, $\theta = x^2$, $\phi = x^3$ (see Figure 1.5) write the square of the line element and use this expression to calculate $g_{ij}$, $g^{ij}$, $\Gamma_{ijk}$, $\Gamma^k_{ij}$.

5.5. Establish the transformation from cartesian coordinates to the spherical coordinates of the preceding problem. Using this transformation, calculate for the coordinates $x^i$ the following quantities: $\mathbf{g}_i$, $\mathbf{g}^i$, $g_{ij}$, $g^{ij}$, $\Gamma_{ijk}$, $\Gamma^k_{ij}$.

5.6. With reference to a cartesian coordinate system $x^{\alpha'}$ the elastic moduli for plane strain are given in the following form:

$$E^{1'1'1'1'} = A, \qquad E^{1'1'2'2'} = B,$$
$$E^{2'2'2'2'} = C, \qquad E^{1'2'1'2'} = G.$$

All moduli not dependent on these by symmetry relations are zero.

For a polar coordinate system $r = x^1$, $\theta = x^2$ calculate $E^{1122}|_1$. Make a plan how to find this derivative and then calculate only those moduli $E^{ijlm}$ and only those Christoffel symbols which are needed to answer the question.

## References

For collateral study the reader is referred to the books mentioned on page 22.

There is more diversity of notation for the Christoffel symbols than for other quantities and operations occurring in tensor analysis. The oldest notation used for $\Gamma_{ijk}$ and $\Gamma^k_{ij}$ seems to be $[{ij \atop k}]$ and $\{{ij \atop k}\}$. It has practically disappeared from the modern literature. Madelung [20] mentions it, but uses $\Gamma_{k,ij}\,\Gamma^k_{ij}$. Brillouin [3, p. 119], does the same and adds "Les notations Γ de Weyl paraissent plus pratiques, parce qu'elles placent les indices en des positions haut et bas correspondant à nos conventions." Many authors of the last decades, like Synge–Schild [31], Aris [1], Eringen [6], Block [2], Sokolnikoff [30], and Hawkins [14], have tried to combine logic and conservatism and use $[ij,k]$, $\{{k \atop ij}\}$, which, at least in the second symbol, brings the indices into the proper position. Levi-Civita [18] and Wills [35] use $[ij,k]$ and $\{ij,k\}$. Duschek–Hochrainer [5] write $\Gamma_{kij}$, $\Gamma^k_{\cdot ij}$ but mostly use $[ij,k]$ and $\{{k \atop ij}\}$. Coburn [4], Green–Zerna [13], Lass [17], and Schild [28] have the notation adopted in this book.

CHAPTER 6

# The Fundamental Equations of Continuum Mechanics

CONTINUUM MECHANICS draws its equations from three sources of physical information. In Section 4.2 we have already discussed the constitutive equations, which describe the empirical properties of the material. The kinematic relations between the displacement vector and the strain tensor have been explained in Chapter 2, but the treatment given there had a preliminary character because we approached the subject before we had learned how to differentiate a vector. We shall have to come back to this subject. The equilibrium conditions and their dynamic counterpart, the equations of motion, have not yet been touched upon. All this will be done now, and then we shall see how one can condense all these equations to a single one for one surviving unknown.

## 6.1. Kinematic Relations

Equations (2.18) and (2.19) are the linearized and the exact forms of the kinematic equation, but their validity is restricted to coordinate systems in which every component of the metric tensor is a constant, i.e. for which $g_{ij,k} = 0$. Because of the presence of the partial derivatives $u_{i,j}$ etc., these equations are not tensor equations. However, on page 69 we saw that in cartesian coordinates the partial derivatives are identical with the covariant derivatives so that we may replace one by the other. Therefore, we may write

$$\varepsilon_{ij} = \tfrac{1}{2}(u_i|_j + u_j|_i) \tag{6.1}$$

for the linear version of the kinematic equation and

$$\varepsilon_{ij} = \tfrac{1}{2}(u_i|_j + u_j|_i + u^k|_i u_k|_j) \tag{6.2}$$

for the exact equation. In this form they are tensor equations and can be carried over into all coordinate systems.

Comparison of (6.1) with (1.51a) shows that for small displacements the strain tensor is the symmetric part of the tensor of the displacement derivatives $u_i|_j$. According to (3.24) the antimetric part $\frac{1}{2}(u_i|_j - u_j|_i)$ may be associated with a vector

$$-2\omega^k = u_i|_j \epsilon^{ijk}. \tag{6.3}$$

Comparison with (5.31) shows that

$$\omega^k \mathbf{g}_k = \tfrac{1}{2} \operatorname{curl} \mathbf{u}.$$

A geometric interpretation is easily obtained by applying (6.3) in a cartesian coordinate system. In this system,

$$\omega^3 = -\tfrac{1}{2}(u_{1,2}\,\epsilon^{123} + u_{2,1}\epsilon^{213}) = \tfrac{1}{2}(u_{2,1} - u_{1,2}).$$

As may be seen from Figure 6.1, the derivatives $u_{2,1}$ and $-u_{1,2}$ are the angles of rotation of line elements $dx^1$ and $dx^2$, both positive when counterclockwise. The vector component $\omega^3$ is their average and may be considered as representing one component of the average rotation of a deformable volume element.

There are three displacement components $u_i$, but six different strain components $\varepsilon_{ij}$. If we want to use (6.1) to calculate the strains from the displacements, we have one equation for every $\varepsilon_{ij}$, but if we consider them as differential equations for the $u_i$, we have three more equations than unknowns. We cannot expect that these equations are compatible with each other for any given set of six functions $\varepsilon_{ij}(x^k)$. There must exist some conditions to which these functions are subjected.

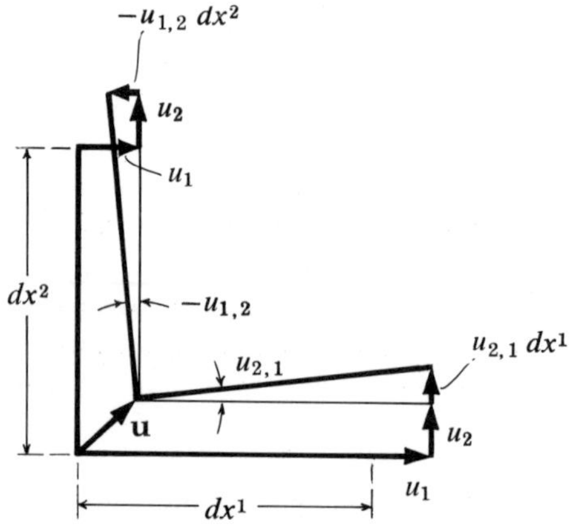


FIGURE 6.1 *Average rotation of a deformable element.*

We find these *compatibility conditions* by eliminating the displacements from (6.1). We calculate the second derivative

$$2\varepsilon_{ij}|_{kl} = u_i|_{jkl} + u_j|_{ikl}$$

and see that the first term on the right is symmetric with respect to the indices $j$ and $l$. Its product with the antimetric permutation tensor $\epsilon^{jln}$ vanishes. Similarly, the product of the second term and $\epsilon^{ikm}$ is zero, and when we multiply the equation with both factors, we have

$$\varepsilon_{ij}|_{kl}\,\epsilon^{ikm}\epsilon^{jln} = 0.$$

This is the compatibility condition. There are six essentially different component equations (6.4) for the choices $mn = 11, 22, 33, 12, 23, 31$. Let us look at two typical samples.

For $m = n = 3$ we have

$$\varepsilon_{11}|_{22}\,\epsilon^{123}\epsilon^{123} + \varepsilon_{12}|_{21}\,\epsilon^{123}\epsilon^{213} + \varepsilon_{21}|_{12}\,\epsilon^{213}\epsilon^{123} + \varepsilon_{22}|_{11}\epsilon^{213}\epsilon^{213} = 0$$

or

$$\varepsilon_{11}|_{22} + \varepsilon_{22}|_{11} - 2\varepsilon_{12}|_{12} = 0.$$

Transcribed into cartesian coordinates $x$, $y$, $z$, this yields the well-known equation

$$\frac{\partial^2 \varepsilon_x}{\partial y^2} + \frac{\partial^2 \varepsilon_y}{\partial x^2} - \frac{\partial^2 \gamma_{xy}}{\partial x\,\partial y} = 0.$$

Now consider a case where $m \neq n$, for example m $= 2$, $n = 3$. Then

$$\varepsilon_{11}|_{32}\,\epsilon^{132}\epsilon^{123} + \varepsilon_{12}|_{31}\,\epsilon^{132}\epsilon^{213} + \varepsilon_{31}|_{12}\,\epsilon^{312}\epsilon^{123} + \varepsilon_{32}|_{11}\,\epsilon^{312}\epsilon^{213} = 0,$$

in cartesian coordinates, after multiplication by a factor 2:

$$2\,\frac{\partial^2 \varepsilon_x}{\partial y\,\partial z} = \frac{\partial}{\partial x}\left[\frac{\partial \gamma_{zx}}{\partial y} + \frac{\partial \gamma_{xy}}{\partial z} - \frac{\partial \gamma_{yz}}{\partial x}\right],$$

and this is also one of a set of three compatibility equation known from the theory of elasticity.

The compatibility conditions belong to the general group of kinematic relations. However, they are not the primary equations of this group, but a result of mathematical operation performed upon (6.1), which is the primary relation.

## 6.2. Condition of Equilibrium and Equation of Motion

We now turn to another group of fundamental equations of continuum mechanics, the conditions of equilibrium.

Figure 6.2 shows an infinitesimal block with parallel faces, cut from a solid or fluid material supposed to be in equilibrium. Its edges are vectors $dr^i\,\mathbf{g}_i$, $ds^j\,\mathbf{g}_j$, $dt^k\,\mathbf{g}_k$. The face on its right-hand side has the area $dA_l = ds^j\,dt^k\epsilon_{jkl}$. Through it and through the face opposite it forces are transmitted which, in first approximation, are $dA_l\,\sigma^{lm}\mathbf{g}_m$ and which, in second approximation, differ by

$$(dA_l\,\sigma^{lm})|_i\,dr^i\,\mathbf{g}_m = dA_l\,\sigma^{lm}|_i\,dr^i\,\mathbf{g}_m = ds^j\,dt^k\,\epsilon_{jkl}\,\sigma^{lm}|_i\,dr^i\,\mathbf{g}_m.$$

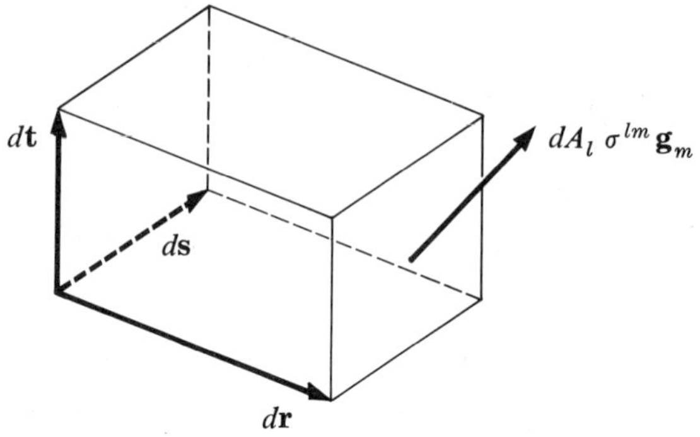


FIGURE 6.2 *Stresses acting on a volume element.*

This is the amount by which the force on the right-hand side is greater than that on the opposite face. The other two pairs of faces of the volume element produce similar unbalanced forces, and all three contributions add up to

$$dr^i\,ds^j\,dt^k(\sigma^{lm}|_i\,\epsilon_{jkl} + \sigma^{lm}|_j\,\epsilon_{kil} + \sigma^{lm}|_k\,\epsilon_{ijl})\mathbf{g}_m. \tag{6.5}$$

We met a similar expression in (5.38), but in that equation we were dealing with a vector $u^l$ instead of the tensor $\sigma^{lm}$. Repeating the argument presented there, we verify that the expression (6.5) equals

$$dr^i\,ds^j\,dt^k\sigma^{lm}|_l\,\epsilon_{ijk}\,\mathbf{g}_m.$$

This is the resultant of all the stresses acting on the volume element. Because of (3.44), it may be written in the simpler form $\sigma^{lm}|_l\,dV\,\mathbf{g}_m$ or, with a change of the dummy indices, $\sigma^{ji}|_j\,dV\,\mathbf{g}_i$. In addition there may be a volume force (mass force), which we write in the form $X^i\,dV\,\mathbf{g}_i$. The sum of the unbalanced stresses and this volume force must be zero:

$$(\sigma^{ji}|_j + X^i)dV\,\mathbf{g}_i = 0.$$

This equation must hold separately for each of the three components, whence, after dropping the constant factor $dV$, we have

$$\sigma^{ji}|_j + X^i = 0. \tag{6.6}$$

This is the condition of equilibrium for the forces acting on a volume element. The equilibrium of moments is taken care of by (4.8), that is, by the symmetry of the stress tensor $\sigma^{ij}$.

If the body is in motion, we have the choice between two formulations of the problem. In the *Lagrangian* or *particle formulation* we attach coordinates (that is, names) to the individual particles and write the dynamic equations in terms of these coordinates, which are moved and deformed with the material. In the *Eulerian* or *field formulation* we use a fixed, rigid coordinate system and let the material move in it. In the first case our equations say what happens to a certain particle, in the second case they say what happens at a certain point in space.

For the *small* motions of a solid both formulations coincide. We can associate a certain, time-dependent velocity with each particle, but since the particle always stays close to the position which it has at rest, the same velocity is also associated with that point in space. The velocity is simply the partial derivative of the displacement **u**,

$$\dot{\mathbf{u}} = \frac{\partial \mathbf{u}}{\partial t}, \qquad \dot{u}_i = \frac{\partial u_i}{\partial t}$$

with the understanding that **u** is a function of time $t$ and of the coordinates $x^j$ attached to the material. The negative product of acceleration and mass density $\rho$ is the d'Alembert inertia force per unit of volume and must be added to the volume force $X^i$ in (6.6). This leads to the dynamic equation,

$$\sigma^{ji}|_j = -X^i + \rho \ddot{u}^i. \tag{6.7}$$

## 6.3. Fundamental Equation of the Theory of Elasticity

We now have all the basic equations which describe the equilibrium and the small motions of an elastic solid, namely, (i) the kinematic relation (6.1), (ii) the equilibrium condition (6.6) or, in its place, the dynamic equation (6.7), and (iii) Hooke's law (4.11) or (4.25). They stand for $6 + 3 + 6 = 15$ component equations and contain as many unknowns, namely, the 6 components of the symmetric strain tensor $\varepsilon_{ij}$, 6 components of the stress tensor $\sigma^{ij}$, and 3 displacement components $u_i$. It is possible and desirable to eliminate the majority of these unknowns between the equations and thus to reduce the rather large system to a set of three differential equations for the displacement components $u_i$. We show this separately for the anisotropic and the isotropic cases.

If the material is anisotropic, we introduce the expressions for the stresses from (4.11) into the equation of motion (6.7) to obtain

$$(E^{ijlm}\varepsilon_{lm})|_j = -X^i + \rho \ddot{u}^i,$$

and now use the kinematic relation (6.1) to express the strain in terms of the displacements:

$$[E^{ijlm}(u_l|_m + u_m|_l)]|_j = -2X^i + 2\rho\ddot{u}^i. \tag{6.8}$$

This tensor equation, valid for $i = 1, 2, 3$, represents three component equations for the $u_i$ and is called the *fundamental equation* of the theory of elasticity. In a homogeneous material the covariant derivative of the elastic modulus vanishes and we may pull $E^{ijlm}$ before the differential operator:

$$E^{ijlm}(u_l|_m + u_m|_l)|_j = -2X^i + 2\rho\ddot{u}^i. \tag{6.9}$$

In the isotropic case we use (4.25), in which we must raise the index $j$. This is done by multiplying the equation by $g^{jk}$ and, after $j$ has disappeared in the process, replacing $k$ by $j$:

$$\sigma^{ij} = \frac{E}{1+\nu}\left(\varepsilon^{ij} + \frac{\nu}{1-2\nu}\,\varepsilon^m_m\, g^{ij}\right).$$

When this is introduced in (6.7), the following equation results:

$$\frac{E}{(1+\nu)(1-2\nu)}\,[(1-2\nu)\varepsilon^{ij}|_j + \nu\varepsilon^m_m|_j\, g^{ij}] = -X^i + \rho\ddot{u}^i. \tag{6.10}$$

In order to combine this with the kinematic relation (6.1), we must there raise one or both subscripts:

$$\varepsilon^{ij} = \tfrac{1}{2}(u^i|^j + u^j|^i)$$

and

$$\varepsilon^i_j = \tfrac{1}{2}(u^i|_j + u_j|^i),$$

whence

$$\varepsilon^m_m = \tfrac{1}{2}(u^m|_m + u_m|^m) = u^m|_m\,. \tag{6.11}$$

The results are used on the left-hand side of (6.10), which then reads

$$\frac{E}{(1+\nu)(1-2\nu)}\left[\frac{1-2\nu}{2}(u^i|^j_j + u^j|^i_j) + \nu u^m|_{mj}\, g^{ij}\right]$$

$$= \frac{E}{2(1+\nu)(1-2\nu)}\,[(1-2\nu)u^i|^j_j + u^j|^i_j].$$

Thus we arrive at the following fundamental equation:

$$\frac{E}{2(1+\nu)(1-2\nu)}\,[(1-2\nu)u^i|^j_j + u^j|^i_j] = -X^i + \rho\ddot{u}^i, \tag{6.12}$$

which again stands for three component equations. Equations (6.8) and (6.12) are two forms of the fundamental equation of elastodynamics. If the term with $\ddot{u}^i$ is dropped, they are valid for problems of elastic equilibrium.

It is, of course, possible to derive (6.12) from the more general equation (6.8) by introducing there the expression (4.19b) for the elastic modulus.

For the discussion of further details we drop the body force $X^i$ and introduce the Lamé moduli $\lambda, \mu$ from (4.17). Instead of (6.12) we then have

$$\mu u^i|_j^j + (\lambda + \mu) u^j|_j^i - \rho \ddot{u}^i = 0. \tag{6.13}$$

We apply the divergence operator (5.30) to this equation and then interchange the dummies $i$ and $j$ in the second term:

$$\mu u^i|_{ji}^j + (\lambda + \mu) u^i|_{ij}^j - \rho \ddot{u}^i|_i = 0.$$

Using the Laplace operator from (5.32), we may write our equation in the form

$$\nabla^2 u^i|_i - c^2 \ddot{u}^i|_i = 0, \tag{6.14}$$

where $c^2$ is not a tensor quantity, but the square of a quantity

$$c = \sqrt{\frac{\rho}{\lambda + 2\mu}}, \tag{6.15}$$

which has the dimension of a velocity.

Another important equation is obtained when (6.13) is subjected to the curl operator (5.31):

$$\mu u^i|_j^{jk} \epsilon_{kil} + (\lambda + \mu) u^j|_j^{ik} \epsilon_{kil} - \rho \ddot{u}^i|^k \epsilon_{kil} = 0.$$

In this equation the second term vanishes because $u^j|_j^{ik} = u^j|_j^{ki}$ and the equation assumes the form

$$(\nabla^2 u^i|^k - c^2 \ddot{u}^i|^k) \epsilon_{kil} = 0 \tag{6.16}$$

with

$$c = \sqrt{\frac{\rho}{\mu}}. \tag{6.17}$$

Every solution of the fundamental equation (6.13) must satisfy (6.14) and (6.16), which have been derived from it. Both these equations are homogeneous and have trivial solutions.

Equation (6.16) has the trivial solution

$$u^i|^k \epsilon_{kil} = 0, \qquad \text{i.e.} \qquad \text{curl}\, \mathbf{u} = 0. \tag{6.18}$$

A vector field satisfying this condition may be written as a gradient field

$$\mathbf{u} = \text{grad}\, \phi, \qquad u^i = \phi|^i, \tag{6.19}$$

where $\phi$ is the displacement potential. In a gradient field we have

$$u^i|^j = \phi|^{ij} = \phi|^{ji} = u^j|^i,$$

and when we introduce this in (6.13), it simplifies and reads

$$(\lambda + 2\mu)u^i|_j^j - \rho\ddot{u}^i = 0$$

or

$$\nabla^2 u^i - c^2 \ddot{u}^i = 0 \tag{6.20}$$

with $c$ from (6.15).

On the other hand, (6.14) has the trivial solution

$$u^i|_i = 0, \qquad \text{i.e.} \qquad \operatorname{div} \mathbf{u} = 0. \tag{6.21}$$

Then the second term of (6.13) drops out and the equation again assumes the form (6.20), but with $c$ from (6.17).

To get a physical understanding of the phenomena represented by the two versions of (6.20), we write it in cartesian coordinates $x$, $y$, $z$ and for the corresponding displacement components $u_i = u^i$:

$$\frac{\partial^2 u_i}{\partial x^2} + \frac{\partial^2 u_i}{\partial y^2} + \frac{\partial^2 u_i}{\partial z^2} - c^2 \frac{\partial^2 u_i}{\partial t^2} = 0, \qquad (i = 1, 2, 3). \tag{6.22}$$

These equations are satisfied by

$$u_i = A_i f_i(x - ct), \qquad (i = 1, 2, 3), \tag{6.23}$$

with three arbitrary functions $f_i$. This solution represents a stress system which moves in positive $x$-direction with the speed $c$ through the elastic body, i.e. an elastic wave.

We may either choose $c$ from (6.15) and satisfy (6.18), or we may choose $c$ from (6.17) and then we must satisfy (6.21).

Equation (6.18) is satisfied if we let $A_2 = A_3 = 0$. Then only $u_1 \neq 0$ and $\varepsilon_{11} = \varepsilon_x = \partial u/\partial x$ is the only nonvanishing strain component. This leads to an elastic dilatation

$$\varepsilon_m^m = \varepsilon_x + \varepsilon_y + \varepsilon_z,$$

but there are no shear stresses. Such waves are called *dilatational waves.*

On the other hand, we may satisfy (6.21) by choosing $A_1 = A_3 = 0$. The only nonvanishing displacement is then

$$u_2 = A_2 f_2(x - ct),$$

which is at right angles with the direction in which the wave propagates. The only strain is a shear strain

$$\varepsilon_{12} = \frac{1}{2}\frac{\partial u_2}{\partial x}$$

and the dilatation is zero. These waves are called *shear waves.*

Generalizing from this experience, we conclude that solutions of (6.20) with curl $\mathbf{u} = 0$ are dilatational waves and propagate with the speed $c$ given by (6.15) and that solutions with div $\mathbf{u} = 0$ are shear waves, whose speed of propagation is given by (6.17).

## 6.4. Flow of Viscous Fluids

For the description of fluid flows the Eulerian formulation is generally preferable. Besides the coordinate system $x^j$, which is tied to the particles and is deformed with the fluid, we introduce a rigid coordinate system $y^i$, in which we ultimately want to write our equations. While the same particle has always the same coordinates $x^j$, its change of position is described by the time dependence of its coordinates $y^i$. When we write partial derivatives with respect to time, we must, in each case, specify what is to be held constant, $x^j$ or $y^i$. In this section we shall use the notation $\partial(\ )/\partial t = (\dot{\ })$ when $y^i = \text{const}$ and $D(\ )/Dt$ when $x^j = \text{const}$. In the first case we are measuring the change occurring at a fixed point in space ("local" time derivative), while in the second case we are following a particle on its way ("particle" time derivative).

The change of the coordinate $y^i$ of a particle is identical with its displacement $u^i$ and in particular the infinitesimal displacement $du^i$, which takes place in a time element $dt$, is identical with the increment $dy^i$. We may, therefore, write the velocity of a particle in the two forms

$$v^i = \frac{Du^i}{Dt} = \frac{Dy^i}{Dt}. \tag{6.24}$$

The two time derivatives of any field quantity $p$ (scalar, vector, tensor) are connected through the relation

$$\frac{Dp}{Dt} = \frac{\partial p}{\partial t} + \frac{\partial p}{\partial y^i}\frac{Dy_i}{Dt} = \frac{\partial p}{\partial t} + v^i\frac{\partial p}{\partial y^i}. \tag{6.25}$$

We want to write our final equations in terms of $y^i$ and, therefore, use base vectors $\mathbf{g}_i = \partial\mathbf{r}/\partial y^i$ and derive from them the metric tensor $g_{ij}$, the permutation tensor $\epsilon_{ijk}$, and the Christoffel symbols. All these quantities are time-independent, while the metric based upon the deformable coordinates $x^j$ is changing with time.

We have already introduced the dot notation for the time derivative at constant $y^i$. We shall use the comma notation for partial derivatives with respect to $y^i$ and the slash notation for covariant derivatives with respect to these coordinates:

$$p_{,i} = \frac{\partial p}{\partial y^i}, \qquad \mathbf{v}_{,i} = \frac{\partial \mathbf{v}}{\partial y^i} = (v^j{}_{,i} + v^k\Gamma^j_{ik})\mathbf{g}_j = v^j|_i\,\mathbf{g}_j. \tag{6.26}$$

We may then write (6.25) in the form

$$\frac{Dp}{Dt} = \dot{p} + v^i p_{,i}. \tag{6.27}$$

The acceleration is the time derivative of the velocity for constant $x^j$:

$$\mathbf{a} = \frac{D\mathbf{v}}{Dt} = \dot{\mathbf{v}} + v^j \mathbf{v}_{,j},$$

in components:

$$a^i = \frac{Dv^i}{Dt} = \dot{v}^i + v^j v^i|_j. \tag{6.28}$$

The formula is easily interpreted. When the flow is not stationary, the velocity at a fixed location changes with time and $\dot{v}^i$ measures the rate of this change. When the flow is stationary, $\dot{v}^i = 0$; but even then the velocity of a particle changes because in the time $dt$ it moves to another point of the flow field where the velocity is different. The amount of acceleration stemming from this cause depends on the space derivative of the velocity and on how far the particle moves in the time $dt$, hence on the product of $v^i|_j$ and $v^j$.

The dynamic equation of fluid flow is the same as that used for solids, (6.7), but we avoid the use of displacements and replace $\ddot{u}^i$ by $a^i$, for which we use (6.28). Hence,

$$\sigma^{ji}|_j = -X^i + \rho(\dot{v}^i + v^j v^i|_j). \tag{6.29}$$

When we differentiate the kinematic relation (6.1) with respect to time for $x^j = \text{const}$, we arrive at the strain rate

$$\frac{D\varepsilon_{ij}}{Dt} = \frac{1}{2}(v_i|_j + v_j|_i). \tag{6.30}$$

In view of the constitutive equations to be used, we split this into a rate of dilatation and a rate of distortion according to (4.29) and (4.30). Using also (6.11), we find that

$$3\frac{De}{Dt} = \frac{D}{Dt}\varepsilon^m_m = v^m|_m \tag{6.31a}$$

and

$$\frac{De^i_j}{Dt} = g^{ik}\frac{D\varepsilon_{jk}}{Dt} - \delta^i_j \frac{De}{Dt} = \frac{1}{2}(v_j|_k + v_k|_j)g^{ik} - \frac{1}{3}v^m|_m \delta^i_j,$$

whence finally

$$\frac{De^i_j}{Dt} = \frac{1}{2}(v_j|^i + v^i|_j) - \frac{1}{3}v^m|_m \delta^i_j. \tag{6.31b}$$

Here we shall consider a viscous, incompressible fluid. For this material, (4.34a) degenerates into the kinematic statement $De/Dt = 0$, whence

$$v^m|_m = 0. \tag{6.32}$$

This incompressibility condition is also known as the continuity condition of the flow field. It states that the divergence of the velocity field is zero. Equation (6.31b) simplifies in this case and there is no difference between $e^i_j$ and $\varepsilon^i_j$:

$$\frac{De^i_j}{Dt} = \frac{D\varepsilon^i_j}{Dt} = \frac{1}{2}(v_j|^i + v^i|_j). \tag{6.33}$$

In a fluid at rest and also in an inviscid fluid in motion, the pressure $p$ is the same in all directions. When a viscous fluid is in motion, the volume elements are undergoing a deformation and there are compressive stresses of different magnitude in different directions. If we still want to speak of a pressure in the fluid, we may define it as the average of the normal stress, to be counted positive when compressive,

$$p = -s = -\tfrac{1}{3}\sigma^m_m. \tag{6.34}$$

When this is introduced in (4.30b), it reads

$$s^i_j = \sigma^i_j + p\delta^i_j,$$

and this may be used in the constitutive relation (4.34b), in which the time derivatives are, of course, particle derivatives:

$$\sigma^i_j + p\delta^i_j = 2\mu\frac{De^i_j}{Dt}. \tag{6.35}$$

With the help of the kinematic relation (6.33) we bring this into the form

$$\sigma^i_j = -p\delta^i_j + \mu(v_j|^i + v^i|_j).$$

We replace $j$ by $k$ and then multiply by $g^{jk}$ to obtain a similar expression in the contravariant components:

$$\sigma^{ij} = -pg^{ij} + \mu(v^j|^i + v^i|^j). \tag{6.36}$$

This equation combines the information contained in the kinematic relation and in one of the constitutive equations (4.34). Our last step on the way to a fundamental equation is the elimination of $\sigma^{ij}$ between (6.29) and (6.36). We find that

$$-p|_j g^{ij} + \mu(v^j|^i_j + v^i|^j_j) = -X^i + \rho(\dot{v}^i + v^j v^i|_j).$$

On the left-hand side, $v^j|^i_j$ is the $i$-derivative of $v^j|_j$, which, from (6.32), vanishes identically. Rearranging the remaining terms, we bring the equation in the following form:

$$\rho\dot{v}^i + \rho v^j v^i|_j - \mu v^i|_j^j = X^i - p|^i. \tag{6.37}$$

This is the Navier–Stokes equation. It stands for three component equations and contains four unknowns, $v^i$ and $p$. Together with the continuity condition (6.32) it determines the flow of the fluid.

In the velocity field $v^i$ we may conceive a set of curves which are everywhere tangent to the vector $\mathbf{v}$. They are called the *streamlines* of the flow field. Similar lines are possible in any other vector field. Magnetic and electric field lines are well-known examples. When we draw the streamlines that pass through all the points of a closed curve $C$, they form a tube called a stream tube. Now let us choose a special coordinate system $y^i$ in which the base vector $\mathbf{g}_1$ has the same direction, though not necessarily the same magnitude, as the velocity vector $\mathbf{v}$. Then $v = v^1\mathbf{g}_1$ and $v^2 = v^3 = 0$. For the curve $C$ we choose a quadrilateral with sides $d\mathbf{r} = dr^2\,\mathbf{g}_2$ and $d\mathbf{s} = ds^3\,\mathbf{g}_3$. The area enclosed by $C$ is then

$$d\mathbf{A} = d\mathbf{r} \times d\mathbf{s} = dr^2\,ds^3\,\epsilon_{231}\mathbf{g}^1$$

and the flux of fluid through it is

$$\mathbf{v} \cdot d\mathbf{A} = v^1\,dr^2\,ds^3\,\epsilon_{231}.$$

When we intersect the stream tube at some other place with a surface $y^1 = \text{const}$, the cross section is again the cross product of some vectors $dr^2\,\mathbf{g}_2$ and $ds^3\,\mathbf{g}_3$ and since the wall of the stream tube is formed of lines along which only $y^1$ varies, the components $dr^2$ and $ds^3$ of these vectors are the same as for $C$, although the base vectors may be different. Therefore, the rate of change of the flux from one cross section of the stream tube to an adjacent one is

$$(\mathbf{v} \cdot d\mathbf{A})_{,1} = dr^2\,ds^3(v^1\epsilon_{231})_{,1} = dr^2\,ds^3(v^1\epsilon_{231})|_1 = dr^2\,ds^3\,\epsilon_{231}v^1|_1$$

and this vanishes because of the continuity condition (6.32). We thus arrive at the result that in a unit of time the same amount of fluid passes through all cross sections of a stream tube. For the velocity field of an incompressible fluid this is physically obvious, but the same mathematical reasoning applies to all vector fields which satisfy (6.32), that is, whose divergence vanishes. We shall soon make use of this fact in a vector field for which the result is not physically obvious.

If the fluid flow is stationary ($\dot{v}^i = 0$), the streamlines are the paths followed by fluid particles during their motion in the flow field. If the flow is not stationary, the shape of the streamlines changes with time and an individual particle follows only one line element of a streamline and continues to follow a line element of a slightly changed streamline and so on. The pathlines $x^j = \text{const}$ are then different from the streamlines.

When the fluid is at rest, the velocity terms must be dropped from (6.37) and all that remains is the statement

$$p|_i = X_i, \tag{6.38}$$

which determines the hydrostatic pressure. Since $p$ is a scalar, $X_i$ must be a gradient field,

$$X_i = \Omega|_i, \tag{6.39}$$

and, hence, must satisfy the condition

$$\text{curl}\,\mathbf{X} = 0, \qquad X_j|_i\,\epsilon^{ijk} = 0. \tag{6.40}$$

Otherwise, the fluid cannot be in equilibrium.

In many important fluids, like water and air, the viscosity $\mu$ is small and can often be disregarded. In such inviscid flow the Navier–Stokes equation (6.37) loses the viscosity term and reads as follows:

$$\rho\dot{v}_i + \rho v^j v_i|_j = X_i - p|_i. \tag{6.41}$$

Comparison of the left-hand side of this equation with (6.28) shows that it equals $\rho(Dv_i/Dt)$.

Now let us consider a closed space curve $C$ in the velocity field, its points attached to the fluid particles, and calculate the time derivative of the circulation integral (see p. 81) of the velocity:

$$\frac{D}{Dt}\oint \mathbf{v}\cdot d\mathbf{r} = \frac{D}{Dt}\oint v_i\,dy^i = \oint \frac{Dv_i}{Dt}\,dy^i + \oint v_i\,\frac{D\,dy^i}{Dt}.$$

The line element $d\mathbf{r} = dy^i\,\mathbf{g}_i$ is the difference of the position vectors pointing from a fixpoint to two adjacent particles on $C$. We may interchange the two differentiations in the last integral and then use (6.24) to obtain

$$\oint v_i\,\frac{D\,dy^i}{Dt} = \oint v_i\,d\,\frac{Dy^i}{Dt} = \oint v_i\,dv^i = \frac{1}{2}\oint (v_i\,dv^i + dv_i\,v^i) = \frac{1}{2}\,[v_i v^i].$$

The bracketed term represents the difference of the values $v_i v^i$ at both ends of $C$. Since the curve is closed, its ends coincide and the difference equals zero. This leaves us with

$$\frac{D}{Dt}\oint v_i\,dy^i = \oint \frac{Dv_i}{Dt}\,dy^i = \frac{1}{\rho}\oint (X_i - p|_i)\,dy^i,$$

which, with (5.19) and (6.39), yields

$$\frac{1}{\rho}\oint (\Omega_{,i} - p_{,i})\,dy^i = \frac{1}{\rho}\,[\Omega - p],$$

and also this difference of terminal values equals zero for a closed curve $C$. Thus we arrive at the statement that for an inviscid fluid moving in a conservative force field

$$\oint v_i \, dy^i = \text{const} \tag{6.42}$$

for any closed curve moving with the particles of the fluid. This is in particular true for any infinitesimal curve bounding an area element $d\mathbf{A}$. From Stokes' theorem (5.42) we conclude that then curl $\mathbf{v} \cdot d\mathbf{A}$ is also a constant. Since, in a general flow field, $d\mathbf{A}$ changes magnitude and direction, this does not mean that the vorticity

$$\boldsymbol{\omega} = \text{curl } \mathbf{v}$$

is a constant.

Also the vorticity vector $\boldsymbol{\omega}$ forms a vector field and we may define a kind of streamlines, which we shall call vorticity lines or vortex filaments and we may consider a vortex tube formed by the vortex filaments passing through all the points of a closed curve $C$. The product $\boldsymbol{\omega} \cdot d\mathbf{A}$ is the flux of vorticity through the cross section $d\mathbf{A}$ of such a tube. Since we saw on page 76 that every curl field has a vanishing divergence, we may repeat the reasoning given for the stream tubes and conclude that along a vortex tube, for constant time, the flux $\boldsymbol{\omega} \cdot d\mathbf{A}$ is constant. The flux through the wall of the vortex tube is zero by its definition. When we now attach the points of this tube to the fluid particles, we see that the flow through any element of its wall will remain zero and that these particles will for all times be the wall of a vortex tube. We also see that a cross section of the tube, no matter how it may change, will always remain a cross section of this tube and will have the same flux $\boldsymbol{\omega} \cdot d\mathbf{A}$. The volume integral of the vorticity vector $\boldsymbol{\omega}$ extended over an element of a vortex tube will always be the same and will stay with the same particles.

Many unsteady flows start from rest and in many steady flows the fluid approaches from infinity with a uniform velocity toward an obstacle creating a nontrivial velocity field. In both cases there was originally curl $\mathbf{v} = 0$ for each fluid particle, and since the dot product of the curl with any area element is constant, it follows that in these cases curl $\mathbf{v} = 0$ throughout and that, according to (5.45), the flow field is a gradient field. The velocity may be written as the gradient of a scalar $\Phi$, called the velocity potential:

$$v_i = \Phi|_i = \Phi_{,i}. \tag{6.43}$$

A velocity field of this kind is called a *potential flow*. Introducing (6.43) into the continuity condition (6.32), one obtains a differential equation for the flow potential:

$$\Phi|_i^i = 0, \tag{6.44}$$

the Laplace equation. The introduction of (6.43) into (6.41) yields an equation for the pressure $p$:

$$p|_i = X_i - \rho(\dot{\Phi}|_i + \Phi|^j\Phi|_{ij}) = X_i - \rho(\dot{\Phi} + \tfrac{1}{2}\Phi|^j\Phi|_j)|_i. \qquad (6.45)$$

The boundary condition for the differential equation (6.45) is a statement regarding the velocity component normal to the boundary $S$ of the domain in which fluid flow takes place. Let $n^i$ be a unit vector normal to the surface $S$, then the normal component of the velocity is

$$v_i n^i = \Phi|_i n^i.$$

It may not be prescribed entirely arbitrarily, but is subject to an integrability condition imposed by the divergence theorem (5.40). When we apply this equation to the velocity field $v^i$, then, because of the continuity condition (6.32), the integrand on the left-hand side vanishes identically and the boundary values of $v_i n^i$ must be so chosen that

$$\int_S v^i \, dA_i = 0.$$

## 6.5. Seepage Flow

The fluid flow considered thus far is the flow of a mass of fluid filling a certain part of space. We now turn our attention to a quite different kind of flow, the seepage flow of a fluid through a porous medium.

There are two kinds of porous materials. In those made to be used as floats the pores are closed cavities, not connected with each other and not accessible from the outside. In these materials no flow is possible. Other materials, such as sand, porous rocks, concrete, filter materials, etc., have a system of communicating pores and a fluid may flow through a maze of tiny channels, bifurcating and recombining randomly. In most cases the flow channels are so small that the movement of the fluid through them is dominated by viscous friction and its velocity is proportional to the pressure gradient.

Since the flow channels are very irregular, the actual velocity of the fluid varies within a wide range and nothing is known about the details. The velocity which we shall use in our equations is not any average of the actual velocities, but is based on the observable mass flow through the bulk volume of the porous medium. To get a precise definition of this velocity, we consider the tetrahedral element shown in Figure 6.3. This element has the same shape as the one shown in Figure 4.1 and the geometric statements made there may be used. In particular, the outer normal of the triangle $ABC$ is

$$d\mathbf{A} = dA_i \, \mathbf{g}^i,$$

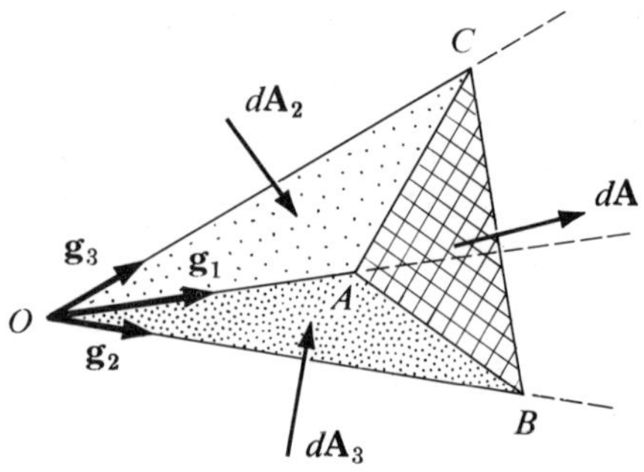


FIGURE 6.3 *Definition of the filter velocity.*

where $dA_1\,\mathbf{g}^1$, $dA_2\,\mathbf{g}^2$, and $dA_3\,\mathbf{g}^3$ are the inner normals of the triangles $OBC$, $OCA$, and $OAB$, respectively. If the tetrahedron is small enough, the volume of fluid which enters or leaves it per unit of time through any one of its sides is proportional to the area of that side. Let the fluid entering through $OBC$ be $v^1\,dA_1$. Similar amounts enter through $OCA$ and $OAB$, and the sum of all three is

$$v^i\,dA_i = v^j\delta^i_j\,dA_i = v^j\mathbf{g}_j\cdot\mathbf{g}^i\,dA_i = \mathbf{v}\cdot d\mathbf{A}.$$

The same amount of fluid must leave through the face $ABC$. The vector

$$\mathbf{v} = v^j\mathbf{g}_j$$

is the velocity of the fluid flow with respect to the bulk cross section. It has been called the *filter velocity*. If the area element $ABC$ is at right angles with $\mathbf{v}$, then its normal vector $d\mathbf{A}$ and the velocity vector $\mathbf{v}$ have the same direction and the flux through the element is the product of $|\mathbf{v}|$ and the total area, solid and pores, of that side of the tetrahedron.

The flow through the pores of a solid body is a viscous flow and no flow will take place unless there is a force field to produce and maintain it. If a certain piece of a porous medium is saturated with fluid and so enclosed that the fluid cannot move, there will be a hydrostatic pressure field satisfying (6.38), where $X_i$ is the force *acting on a unit of the fluid volume* (not of the bulk volume of the porous medium). The vector $X_i$ may simply be the specific weight of the fluid referred to the base vectors $\mathbf{g}^i$, or an electrostatic force, or a d'Alembert inertia force like the centrifugal force acting on the fluid in a rotating filter.

To make the fluid move, $p|_i - X_i$ must be different from zero and if the pore geometry has no preferential direction, the filter velocity follows the direction of the force field $X_i - p|_i$ and is proportional to its intensity:

$$v_i = k(X_i - p|_i) = k(\Omega - p)|_i. \qquad (6.46)$$

This linear law is known as Darcy's law. Some porous media are anisotropic. The pores may be arranged in long strings going all in one direction, or the material may have layers, in which flow is easy, while flow at right angles to these layers meets with much more resistance. To cover such cases, we replace the permeability coefficient $k$ by the permeability tensor $k^{ij}$ and write

$$v^i = k^{ij}(X_j - p|_j) = k^{ij}(\Omega - p)|_j. \tag{6.47}$$

Continuity of the flow again requires that the divergence of the velocity field be zero,

$$v^i|_i = 0, \tag{6.48}$$

and when this and (6.47) are combined, a differential equation for the pressure $p$ results:

$$(k^{ij}p|_j)|_i = (k^{ij}X_j)|_i. \tag{6.49}$$

In a homogeneous medium $k^{ij}|_i = 0$ (see p. 72), and then we have simply

$$k^{ij}p|_{ij} = k^{ij}X_j|_i, \tag{6.50}$$

which in the isotropic case $k^{ij} = kg^{ij}$ simplifies to

$$kg^{ij}p|_{ij} = kg^{ij}X_j|_i,$$

whence

$$p|_i^i = X^i|_i. \tag{6.51}$$

This is the Poisson equation and, in the absence of body forces, the Laplace equation.

There seems to be no fundamental physical law which would require that the permeability tensor be symmetric. Proofs of symmetry may be found in the literature [7], [27], but these are based on a somewhat vague idea of flow channels in a random arrangement and are not fully convincing. The author has tried to find a counterexample and has studied several regular, two-dimensional pore systems, which were complicated enough to appear promising. In all cases the tensor $k^{\alpha\beta}$ turned out to be symmetric. Of course, the negative result of this effort is not a conclusive proof of the symmetry, but unless there is a valid demonstration to the contrary, it seems to be safe to assume that $k^{ij} = k^{ji}$.

The fluid moving through the channels of a porous medium is exerting pressure and friction forces on the solid matrix, which act on it as an external load producing stresses. Before we can study these stresses, we need a geometric concept, the *porosity*.

Figure 6.4a shows the same element as Figure 6.2, but embedded in a cartesian coordinate system $x, y, z$. The dark part of its surface represents

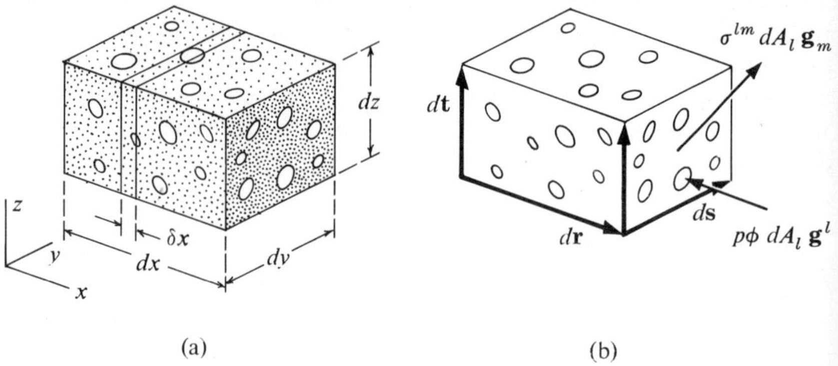


FIGURE 6.4 *Volume element of a porous material.*

solid matter while the small circles are sections through pores. When we cut the material by several parallel planes, the share of the pores in the total cross section vacillates around an average, in which alone we are interested. Let the section $dy\,dz$ on the right be an average section and let $\phi\,dy\,dz$ be the area of the pores and $(1-\phi)\,dy\,dz$ the solid area of the section.

The dimensions $dx$, $dy$, $dz$ of the volume element are, as in Figure 6.3, infinitesimal in the sense that they are very small compared to the dimensions of the flow field, but they are still large compared to the dimensions of the individual pores. We consider now a slice whose thickness $\delta x$ is of the order of the pore diameter or even smaller. In the statistical average, the solid volume in this slice is $(1-\phi)\,dy\,dz\,\delta x$. When we add the volume of all the slices which constitute the element $dx\,dy\,dz$, we find that its solid volume is $(1-\phi)\,dx\,dy\,dz$ and its pore volume $\phi\,dx\,dy\,dz$. Thus the same quantity $\phi$, called the *porosity* of the material, describes the amount of pores in a volume and in a cross section. From this fact it follows that $\phi$ must be the same for sections in any direction, even if the arrangement of the pores has a preferred direction.

It is worthwhile to draw attention here to a few details of pore geometry.

(i) The pores whose cross sections appear in Figure 6.4a are not isolated cavities, but parts of a system of coherent channels. In a material with isolated cavities no seepage flow is possible.

(ii) If a material contains both pore channels and closed bubbles, the latter ones count for our purposes as part of the solid matrix.

(iii) The figure represents a spongelike material. In a gravel bed, the pores are the coherent part of the section and the solid particles appear as islands. Nevertheless, the definition of $\phi$ and all later statements including the equilibrium condition (6.53) are equally valid.

Now let us turn to Figure 6.4b, which shows the same element, but with notations appropriate for the use of general, curvilinear coordinates $x^i$. On the side

$$d\mathbf{s} \times d\mathbf{t} = ds^i\, dt^k\, \epsilon_{ikl}\, \mathbf{g}^l = dA_l\, \mathbf{g}^l$$

two forces are acting, the resultants of the stress $\sigma^{lm}$ in the solid matrix and of the fluid pressure $p$ in the pores. It is convenient (and unobjectionable) to define the stress in terms of the gross cross section so that the force is $\sigma^{lm}\, dA_l\, \mathbf{g}_m$, as indicated in the figure. The fluid pressure $p$ acts on the cross section $\phi\, dA_l\, \mathbf{g}^l$ of the pores and is opposite in direction to the outer normal of the section. Therefore, the total force acting on this side of the volume element is

$$d\mathbf{F} = \sigma^{lm}\, dA_l\, \mathbf{g}_m - p\phi\, dA_l\, \mathbf{g}^l = (\sigma^{lm} - p\phi g^{lm})\, dA_l\, \mathbf{g}_m. \tag{6.52}$$

This force takes the place of the force $\sigma^{lm}\, dA_l\, \mathbf{g}_m$ used in the derivation of (6.5). When we introduce it there and then follow the derivation of the equilibrium condition (6.6), we find in its place the equation

$$(\sigma^{ij} - p\phi g^{ij})|_j + X^i = 0,$$

which we may write in the form

$$\sigma^{ij}|_j = -X^i + (p\phi)|^i. \tag{6.53}$$

The body force $X^i$ is the sum of all actual body forces (gravity, centrifugal force, etc.) acting on the solid and fluid material in a volume element. We see that the presence of a pressure field in the fluid has on the stress system in the solid material the effect of an additional body force. It is remarkable that even a constant pressure $p$ produces stresses $\sigma^{ij}$ if the porosity is variable.

For the formulation of stress problems in a solid subjected to a seepage flow, (6.53) must be combined with Hooke's law (4.25) and the kinematic relation (6.1). Since the measurement of elastic constants is based on the gross dimensions of the test specimen, the use of gross stress $\sigma^{ij}$ is practical and unobjectionable. If the solid matrix breaks under the stress caused by the pressure field $p$, the surface of rupture will deviate from a smooth surface in the sense that it seeks the weakest places in the pore structure, where the walls are thinnest and the pores largest. This, of course, is equally true if a tensile test is performed with a dry specimen and the gross stress of rupture includes this weakness as in any other tensile test. However, in the actual surface of rupture the porosity has a value higher than the average $\phi$. Since the total force $d\mathbf{F}$ transmitted through one of its area elements is the same as in a surface of average character with which it almost coincides, it is seen from (6.52) that in the surface of rupture *even the gross stress*, if tensile, is higher than the value calculated from the differential equations of the problem. Since the porosity in an irregular rupture surface is hard to measure, it appears necessary to determine the strength of the material not in a tension test, but in an actual flow field.

*Problems*

6.1. Express the kinematic relation (6.1) in cylindrical coordinates $x^1 = r$, $x^2 = \theta$, $x^3 = z$ shown in Figure 6.5.

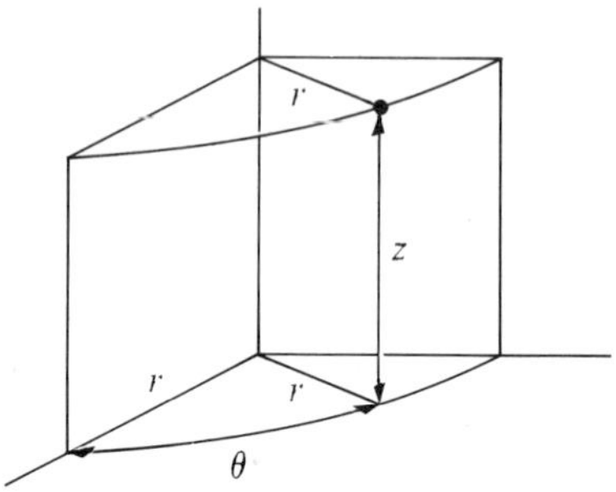


FIGURE 6.5

6.2. Drop the dynamic term from the fundamental equation of elasticity (6.12) and then write this equation as three component equations for the spherical coordinate system shown in Figure 1.5.

6.3. Write the continuity condition (6.32) and the Navier–Stokes equation (6.37) for plane polar coordinates.

**References**

For books on the theory of elasticity see the references on page 65. Cartesian component equations of fluid mechanics are given by Lamb [16] and Milne-Thomson [22], while Aris [1] uses a combination of general tensor notation and Gibbs notation. The flow of a fluid in a nonisotropic medium has been treated by many authors. Most of them think of a stratified rock and take symmetry of the permeability tensor $k^{ij}$ for granted. The subject has been taken up seriously by Ferrandon [7]. His work is described in the book by Scheidegger [27, p. 76]. The force exerted by the fluid upon the solid medium has first been considered as a kind of static buoyancy. A rational formulation was given by Flügge [9] and the problem has been taken up again by Ferrandon [7].

CHAPTER 7

# Special Problems of Elasticity

THE SOLUTION OF the differential equations of the theory of elasticity is greatly simplified when it is known in advance that certain components of the displacement vector $u_i$, the strain tensor $\varepsilon_{ij}$, and the stress tensor $\sigma^{ij}$ are zero or are so small that they may be neglected. It is a historical peculiarity of the literature written in English on the subject that only problems of the first class are considered to belong to the Theory of Elasticity, while those of the second class are relegated to her humble sister, called Strength of Materials. From the point of view of the engineer there is no reason for such a class distinction and we shall accept them as equally honest and equally good attempts to obtain solutions for practicai problems.

## 7.1. Plane Strain

There are two plane problems in the theory of elasticity, the problems of plane strain and of plane stress. In *plane strain*, the actual object is a cylindrical or prismatic body (Figure 7.1), which is so loaded and supported that all stresses and deformations are independent of $z$. We may cut from this body, at any place, a slice of any thickness and study the stress problem in it. The same stresses, strains, and displacements will occur in all such slices. In formulating and solving this problem, we use as coordinates $z = x^3$ and two arbitrary, possibly curvilinear, coordinates $x^\alpha$ in the $x,y$ plane.

In *plane stress* we are dealing with a thin slab with free surfaces, loaded only by forces in its middle plane, that is, in the plane which halves the thickness everywhere. The thickness of the slab is not necessarily constant, but should not vary abruptly. Also in this case we use a general plane coordinate system $x^\alpha$ and a coordinate $x^3$ measured normally to the plane of the other

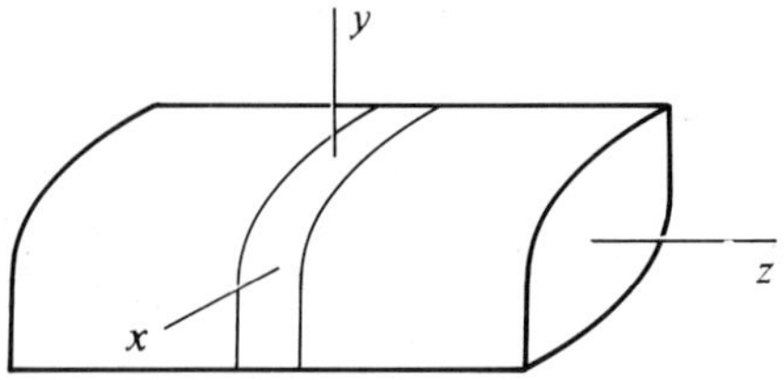


FIGURE 7.1 *Prismatic body in plane strain.*

two, but instead of choosing $x^3 = z$, we may choose $x^3$ such that it assumes the values $\pm 1$ on the free surfaces of the slab. This is particularly advantageous when the slab thickness varies with $x^\alpha$.

The base vectors $\mathbf{g}_\alpha$ of our coordinate system can make any angle with each other, but $\mathbf{g}_3$ is always at right angles to both of them. Borrowing from the terminology of the crystallographers, we may say that we are using a monoclinic reference frame. In this system

$$g_{\alpha 3} = \mathbf{g}_\alpha \cdot \mathbf{g}_3 = 0$$

and the metric tensor has the following components:

$$\begin{bmatrix} g_{11} & g_{12} & 0 \\ g_{21} & g_{22} & 0 \\ 0 & 0 & g_{33} \end{bmatrix}.$$

In the coordinates $x^\alpha, x^3$ some of the general formulas simplify. For example, if we let $k = 3$ in (5.3a) and restrict $i, j$ to the two-dimensional range, we find that

$$\Gamma_{\alpha\beta 3} = \mathbf{g}_{\alpha,\beta} \cdot \mathbf{g}_3 = 0 \tag{7.1a}$$

because $\mathbf{g}_3$ is at right angles to $\mathbf{g}_{\alpha,\beta}$. For a similar reason

$$\Gamma_{\alpha 3\beta} = \Gamma_{3\alpha\beta} = \mathbf{g}_{3,\alpha} \cdot \mathbf{g}_\beta = 0; \tag{7.1b}$$

and also any other $\Gamma_{ijk}$ or $\Gamma^k_{ij}$ vanishes if one or more of its indices are equal to 3.

To write the covariant derivative of a vector

$$\mathbf{v} = v^i \mathbf{g}_i = v^\alpha \mathbf{g}_\alpha + v^3 \mathbf{g}_3,$$

we use (5.11), where we split the sum over $k = 1, 2, 3$ into a sum over $\gamma = 1, 2$ and a separate term for $k = 3$:

$$v^i|_j = v^i{}_{,j} + v^\gamma \Gamma^i_{j\gamma} + v^3 \Gamma^i_{j3}.$$

The last term vanishes. What remains needs to be inspected for different cases. First, let $i = \alpha$, $j = \beta$; then

$$v^\alpha|_\beta = v^\alpha{}_{,\beta} + v^\gamma \Gamma^\alpha_{\beta\gamma}, \tag{7.2a}$$

which is simply (5.11) with a reduced range of all indices. Now let us have $i = \alpha$, but $j = 3$:

$$v^\alpha|_3 = v^\alpha{}_{,3} + v^\gamma \Gamma^\alpha_{3\gamma} = v^\alpha{}_{,3}. \tag{7.2b}$$

Similarly, with $i = 3$ and $j = \alpha$:

$$v^3|_\alpha = v^3{}_{,\alpha} + v^\gamma \Gamma^3_{\alpha\gamma} = v^3{}_{,\alpha} \tag{7.2c}$$

and

$$v^3|_3 = v^3{}_{,3}. \tag{7.2d}$$

In these three cases, the covariant derivative is identical to the common partial derivative. Similar equations hold for the covariant components,

$$v_\alpha|_\beta = v_{\alpha,\beta} - v_\gamma \Gamma^\gamma_{\alpha\beta}, \qquad v_\alpha|_3 = v_{\alpha,3}, \qquad v_3|_\alpha = v_{3,\alpha}, \qquad v_3|_3 = v_{3,3} \tag{7.3}$$

and also for tensors.

In plane strain, which we shall now study, the third displacement component $u_3 = 0$ and the displacement

$$\mathbf{u} = u_\alpha \mathbf{g}^\alpha$$

is a plane vector. The linear kinematic relation (6.1) applies to the in-plane components

$$\varepsilon_{\alpha\beta} = \tfrac{1}{2}(u_\alpha|_\beta + u_\beta|_\alpha), \tag{7.4}$$

while in all other formulas the covariant derivatives are equal to the common derivatives:

$$\varepsilon_{\alpha 3} = \tfrac{1}{2}(u_{\alpha,3} + u_{3,\alpha}) \qquad \text{and} \qquad \varepsilon_{33} = u_{3,3}.$$

In the first of these equations, $u_\alpha$ does not depend on $x^3$ and $u_3 = 0$, and in the second equation again $u_3 = 0$, hence

$$\varepsilon_{\alpha 3} = \varepsilon_{3\alpha} = \varepsilon_{33} = 0. \tag{7.5}$$

For a general, anisotropic material, Hooke's law is derived from (4.11). For the in-plane stresses we write

$$\sigma^{\alpha\beta} = E^{\alpha\beta lm}\varepsilon_{lm} = E^{\alpha\beta\gamma\delta}\varepsilon_{\gamma\delta} + E^{\alpha\beta\gamma 3}\varepsilon_{\gamma 3} + E^{\alpha\beta 3\delta}\varepsilon_{3\delta} + E^{\alpha\beta 33}\varepsilon_{33}.$$

Because of (7.5) the last three terms may be dropped and we have

$$\sigma^{\alpha\beta} = E^{\alpha\beta\gamma\delta}\varepsilon_{\gamma\delta}. \tag{7.6a}$$

In addition, there is a stress

$$\sigma^{33} = E^{33\gamma\delta}\varepsilon_{\gamma\delta}, \tag{7.6b}$$

which is needed to keep $\varepsilon_{33} = 0$, but there might as well be stresses

$$\sigma^{\alpha 3} = E^{\alpha 3\gamma\delta}\varepsilon_{\gamma\delta}, \tag{7.6c}$$

if there are nonzero moduli of this type. We shall look into this matter on page 113 and restrict ourselves at present to the case of plane anisotropy, in which all those moduli vanish which have one or three superscripts 3.

Later there will arise the need to invert (7.6a), that is, to express the strains $\varepsilon_{\gamma\delta}$ in terms of the stresses $\sigma^{\alpha\beta}$. If we take proper account of the symmetry of both these tensors, (7.6a) represents three linear equations for $\alpha\beta = 11$, 12, 22, and with the unknowns $\varepsilon_{11}$, $\varepsilon_{12}$, $\varepsilon_{22}$. They can be solved and the result has the general form

$$\varepsilon_{\gamma\delta} = \bar{C}_{\alpha\beta\gamma\delta}\,\sigma^{\alpha\beta}. \tag{7.7}$$

The quantities $\bar{C}_{\alpha\beta\gamma\delta}$ are called *elastic compliances* and, like the moduli, constitute a plane tensor of the fourth order. A similar inversion is, of course, also possible for the three-dimensional equations (4.11), but it leads to compliances $C_{ijlm}$ which are not identical to $\bar{C}_{\alpha\beta\gamma\delta}$. While the component equations (7.6a) are a cut out of the equations (4.11), the same is not true for the equations (7.7) in relation to their three-dimensional counterpart.

For an isotropic material we start from (4.25), which, with $\varepsilon_3^\alpha = \varepsilon_\alpha^3 = \varepsilon_3^3 = 0$ simply reduces to

$$\sigma_\beta^\alpha = \frac{E}{1+\nu}\left(\varepsilon_\beta^\alpha + \frac{\nu}{1-2\nu}\,\varepsilon_\zeta^\zeta\,\delta_\beta^\alpha\right), \tag{7.8}$$

$$\sigma_3^\alpha = \sigma_\alpha^3 = 0,$$

$$\sigma_3^3 = \frac{E\nu}{(1+\nu)(1-2\nu)}\,\varepsilon_\zeta^\zeta.$$

To invert (7.8), we use the same technique as on page 58. We first let $\beta = \alpha$ and find

$$\sigma_\alpha^\alpha = \frac{E}{1+\nu}\left(\varepsilon_\alpha^\alpha + \frac{2\nu}{1-2\nu}\,\varepsilon_\zeta^\zeta\right) = \frac{E}{(1+\nu)(1-2\nu)}\,\varepsilon_\zeta^\zeta,$$

and then

$$\sigma_\beta^\alpha = \frac{E}{1+\nu}\,\varepsilon_\beta^\alpha + \nu\sigma_\zeta^\zeta\,\delta_\beta^\alpha,$$

whence

$$\varepsilon_\beta^\alpha = \frac{1+\nu}{E}\left(\sigma_\beta^\alpha - \nu\sigma_\zeta^\zeta\,\delta_\beta^\alpha\right). \tag{7.9}$$

Obviously this is not what would be obtained by restricting in (4.27) all terms to the two-dimensional range. One obtains, however, the correct result if he uses the relation between $\sigma_3^3$ and $\varepsilon_\zeta^\zeta$ following (7.8) to express $\sigma_m^m$ in terms of $\sigma_\zeta^\zeta$, that is if he makes use of the side condition $\varepsilon_3^3 = 0$ to eliminate from (4.27) the stress $\sigma_3^3$, the only nonzero quantity which is not part of the plane tensor $\sigma_\beta^\alpha$.

The equilibrium conditions can easily be extracted from (6.6). Since $\sigma^{3\alpha} = 0$, we obtain, for $i = \alpha$,

$$\sigma^{\beta\alpha}|_\beta + X^\alpha = 0, \tag{7.10}$$

and the equation for $i = 3$ has only vanishing terms.

Equations (7.4), (7.10), and either (7.6a), (7.7) or (7.8) are eight component equations for the eight unknowns $\sigma^{11}$, $\sigma^{12}$, $\sigma^{22}$, $\varepsilon_{11}$, $\varepsilon_{12}$, $\varepsilon_{22}$, $u_1$, $u_2$. There are two ways of reducing this system to smaller size by elimination of unknowns.

One way is to eliminate stresses and strains and thus to derive differential equations for the displacement components $u_\alpha$. To do this, we introduce the stresses from (7.6a) into (7.10):

$$(E^{\alpha\beta\gamma\delta}\varepsilon_{\gamma\delta})|_\beta + X^\alpha = 0$$

and then use (7.4) to express the strain in terms of the displacement:

$$[E^{\alpha\beta\gamma\delta}(u_\gamma|_\delta + u_\delta|_\gamma)]|_\beta = -2X^\alpha. \tag{7.11}$$

If the material is homogeneous, the covariant derivative of the modulus vanishes and we have instead

$$E^{\alpha\beta\gamma\delta}(u_\gamma|_{\delta\beta} + u_\delta|_{\gamma\beta}) = -2X^\alpha. \tag{7.12}$$

In the isotropic case we rewrite (7.4) and (7.8) in the form

$$\begin{gathered} \varepsilon_\gamma^\alpha = \tfrac{1}{2}(u^\alpha|_\gamma + u_\gamma|^\alpha), \\ \sigma^{\alpha\beta} = \frac{E}{1+\nu}\left(\varepsilon_\gamma^\alpha g^{\beta\gamma} + \frac{\nu}{1-2\nu}\varepsilon_\zeta^\zeta g^{\alpha\beta}\right) \end{gathered} \tag{7.13}$$

and proceed as before to obtain the following differential equation, which, of course, is a special form of (7.12):

$$u^\alpha|_\beta^\beta + \frac{1}{1-2\nu}u^\beta|_\beta^\alpha + \frac{2(1+\nu)}{E}X^\alpha = 0. \tag{7.14}$$

Each of the tensor equations (7.12) and (7.14) represents a pair of simultaneous differential equations of the second order. Boundary conditions to be imposed upon displacements or stresses can easily be formulated in terms of $u_\alpha$ and $u_\alpha|_\beta$. After the equations have been solved, strains and stresses follow without further integration.

The second way of simplifying the basic equations is more complicated and is not applicable when boundary conditions for the displacements are prescribed. Nevertheless, it enjoys greater popularity (is, in fact, the only one treated in several well-known books on the theory of elasticity), because it leads to one simple equation, which, in the isotropic case, is the bipotential equation.

We begin by postulating that the stress components are the second derivatives of a scalar $\Phi$, the stress function:

$$\sigma^{\alpha\beta} = \epsilon^{\alpha\lambda}\epsilon^{\beta\mu}\Phi|_{\lambda\mu}, \tag{7.15}$$

whence

$$\sigma^{\alpha\beta}|_{\beta} = \epsilon^{\alpha\lambda}\epsilon^{\beta\mu}\Phi|_{\beta\lambda\mu}.$$

There are two such equations, one for $\alpha = 1$ and one for $\alpha = 2$. Once $\alpha$ has been chosen, $\lambda$ also is fixed to make $\epsilon^{\alpha\lambda} \neq 0$. This value $\epsilon^{\alpha\lambda}$ is multiplied by the sum $\epsilon^{\beta\mu}\Phi|_{\beta\lambda\mu}$, which vanishes because of (3.31) and $\Phi|_{\beta\lambda\mu} = \Phi|_{\mu\lambda\beta}$. We see that stresses $\sigma^{\alpha\beta}$ from (7.15) satisfy the equilibrium condition (7.10) with $X^{\alpha} = 0$.

The procedure may easily be generalized to cover the case that the body force $X^{\alpha}$ is conservative and, hence, can be derived from a potential $\Omega$:

$$X^{\alpha} = \Omega|^{\alpha}. \tag{7.16}$$

We replace (7.15) by

$$\sigma^{\alpha\beta} = \epsilon^{\alpha\lambda}\epsilon^{\beta\mu}\Phi|_{\lambda\mu} - g^{\alpha\beta}\Omega \tag{7.17}$$

and have

$$\sigma^{\alpha\beta}|_{\beta} = \epsilon^{\alpha\lambda}\epsilon^{\beta\mu}\Phi|_{\lambda\mu\beta} - g^{\alpha\beta}\Omega|_{\beta}.$$

The first term on the right vanishes again and thus we have

$$\sigma^{\alpha\beta}|_{\beta} = -g^{\alpha\beta}\Omega|_{\beta} = -\Omega|^{\alpha} = -X^{\alpha},$$

which agrees with (7.10). We continue our work using (7.17), which includes, with $\Omega \equiv 0$, the commonly treated case of vanishing body forces.

We need one differential equation for $\Phi$ and have at our disposal the kinematic relation (7.4) and Hooke's law (7.7). We begin with (7.4) and form the expression

$$2\varepsilon_{\gamma\delta}|_{\nu\rho}\,\epsilon^{\gamma\nu}\epsilon^{\delta\rho} = u_{\gamma}|_{\delta\nu\rho}\,\epsilon^{\gamma\nu}\epsilon^{\delta\rho} + u_{\delta}|_{\gamma\rho\nu}\,\epsilon^{\delta\rho}\epsilon^{\gamma\nu}.$$

After a change of notation for the dummy indices in the last term, this may be written

$$2\varepsilon_{\gamma\delta}|_{\nu\rho}\,\epsilon^{\gamma\nu}\epsilon^{\delta\rho} = 2u_{\gamma}|_{\delta\nu\rho}\,\epsilon^{\gamma\nu}\epsilon^{\delta\rho}.$$

For every possible pair $\gamma\nu$ there are two terms in this sum, and because of $u_\gamma|_{\delta\nu\rho} = u_\gamma|_{\rho\nu\delta}$ and (3.31) they add up to zero. Therefore

$$\varepsilon_{\gamma\delta}|_{\nu\rho}\,\epsilon^{\gamma\nu}\epsilon^{\delta\rho} = 0. \tag{7.18}$$

This is the tensor form of the compatibility equation of two-dimensional deformation. It could, of course, have been derived from (6.4) by restricting there the dummy indices $i, j, k, l$ to the two-dimensional range. Then necessarily $m = n = 3$ and the three-dimensional permutation tensor reduces to its two-dimensional counterpart shown in (7.18). This is a second-order differential equation, which the three strain components must satisfy to make it possible to calculate them from the derivatives of only two displacement components $u_\gamma$. We now use the elastic law (7.7) to write this condition in terms of the stresses,

$$(\bar{C}_{\alpha\beta\gamma\delta}\,\sigma^{\alpha\beta})|_{\nu\rho}\,\epsilon^{\gamma\nu}\epsilon^{\delta\rho} = 0,$$

and then express $\sigma^{\alpha\beta}$ in terms of $\Phi$ according to (7.17):

$$(\bar{C}_{\alpha\beta\gamma\delta}\,\Phi|_{\lambda\mu})|_{\nu\rho}\,\epsilon^{\alpha\lambda}\epsilon^{\beta\mu}\epsilon^{\gamma\nu}\epsilon^{\delta\rho} = (\bar{C}_{\alpha\beta\gamma\delta}\,\Omega)|_{\nu\rho}\,g^{\alpha\beta}\epsilon^{\gamma\nu}\epsilon^{\delta\rho}. \tag{7.19}$$

This is the desired differential equation. It is of the fourth order, and as long as all boundary conditions are prescribed in terms of the stresses, they can easily be formulated in terms of second derivatives of $\Phi$. For a material which is homogeneous though still anisotropic, we may again take the compliance $\bar{C}_{\alpha\beta\gamma\delta}$ out of the covariant derivatives and have

$$\bar{C}_{\alpha\beta\gamma\delta}\,\Phi|_{\lambda\mu\nu\rho}\,\epsilon^{\alpha\lambda}\epsilon^{\beta\mu}\epsilon^{\gamma\nu}\epsilon^{\delta\rho} = \bar{C}_{\alpha\beta\gamma\delta}\,\Omega|_{\nu\rho}\,g^{\alpha\beta}\epsilon^{\gamma\nu}\epsilon^{\delta\rho}. \tag{7.20}$$

In the isotropic case we replace (7.7) by (7.9), which we rewrite in the form

$$\varepsilon_{\gamma\delta} = \frac{1+\nu}{E}(\sigma^{\alpha\beta}g_{\alpha\gamma}\,g_{\beta\delta} - \nu\sigma^{\alpha\beta}g_{\alpha\beta}\,g_{\gamma\delta}). \tag{7.21}$$

Introduction of this expression in the compatibility condition (7.18) shows that

$$\sigma^{\alpha\beta}|_{\nu\rho}(g_{\alpha\gamma}\,g_{\beta\delta} - \nu g_{\alpha\beta}\,g_{\gamma\delta})\epsilon^{\gamma\nu}\epsilon^{\delta\rho} = 0$$

and introduction of $\Phi$ from (7.17) yields the differential equation

$$\Phi|_{\lambda\mu\nu\rho}(g_{\alpha\gamma}\,g_{\beta\delta} - \nu g_{\alpha\beta}\,g_{\gamma\delta})\epsilon^{\alpha\lambda}\epsilon^{\beta\mu}\epsilon^{\gamma\nu}\epsilon^{\delta\rho} = g^{\alpha\beta}\Omega|_{\nu\rho}(g_{\alpha\gamma}\,g_{\beta\delta} - \nu g_{\alpha\beta}\,g_{\gamma\delta})\epsilon^{\gamma\nu}\epsilon^{\delta\rho}, \tag{7.22}$$

which, of course, is again of the fourth order. It is still amenable to simplification. We work at the coefficient of $\Phi|_{\lambda\mu\nu\rho}$ and bring it successively into the following forms, making use of (3.29b):

$$\begin{aligned}(\epsilon_\gamma^{\cdot\lambda}\epsilon_\beta^{\cdot\rho} - \nu\epsilon_\beta^{\cdot\lambda}\epsilon_\gamma^{\cdot\rho})\epsilon^{\beta\mu}\epsilon^{\gamma\nu} &= (\epsilon_{\gamma\delta}\,g^{\delta\lambda}\epsilon_{\beta\alpha}\,g^{\alpha\rho} - \nu\epsilon_{\beta\alpha}\,g^{\alpha\lambda}\epsilon_{\gamma\delta}\,g^{\delta\rho})\epsilon^{\beta\mu}\epsilon^{\gamma\nu}\\ &= (g^{\delta\lambda}g^{\alpha\rho} - \nu g^{\alpha\lambda}g^{\delta\rho})\delta_\alpha^\mu\delta_\delta^\nu = g^{\lambda\nu}g^{\mu\rho} - \nu g^{\lambda\mu}g^{\nu\rho}.\end{aligned}$$

When we now multiply this by $\Phi|_{\lambda\mu\nu\rho}$, we interchange the dummies $\mu$ and $\nu$ in the first term. This must, of course, be done in the superscripts of the coefficient and in the subscripts of the derivative, but in the latter it does not change the value and we may restore the former order and have

$$\Phi|_{\lambda\mu\nu\rho}(g^{\lambda\mu}g^{\nu\rho} - \nu g^{\lambda\mu}g^{\nu\rho}) = (1-\nu)\Phi|_{\lambda\nu}^{\lambda\nu}.$$

When the right-hand side of (7.22) is subjected to a similar treatment, it subsequently assumes the following forms:

$$\begin{aligned}\Omega|_{\nu\rho}(\delta_\gamma^\beta \epsilon_\beta^{\cdot\rho} - 2\nu\epsilon_\gamma^{\cdot\rho})\epsilon^{\gamma\nu} &= \Omega|_{\nu\rho}(\epsilon_\gamma^{\cdot\rho} - 2\nu\epsilon_\gamma^{\cdot\rho})\epsilon^{\gamma\nu} \\ &= (1-2\nu)\Omega|_\nu^\rho \epsilon_{\gamma\rho}\epsilon^{\gamma\nu} = (1-2\nu)\Omega|_\nu^\rho \delta_\rho^\nu = (1-2\nu)\Omega|_\rho^\rho.\end{aligned}$$

The differential equation of the problem is now reduced to the following simple form:

$$(1-\nu)\Phi|_{\alpha\beta}^{\alpha\beta} = (1-2\nu)\Omega|_\alpha^\alpha \tag{7.23}$$

and in the case of vanishing body force it is simply

$$\Phi|_{\alpha\beta}^{\alpha\beta} = 0. \tag{7.24}$$

As may be seen from (5.32), this is the tensor form of the two-dimensional bipotential equation $\nabla^2\nabla^2\Phi = 0$.

## 7.2. Plane Stress

This terminates our study of plane strain and we now turn to *plane stress*, restricting ourselves to a slab of constant thickness $h$. On the faces $x^3 = \pm h/2$ there are no external forces, i.e. the stresses $\sigma^{3\alpha} = \sigma^{33} = 0$. Since the slab is thin, we may then expect that these stress components are zero across its thickness, while the strain $\varepsilon_{33}$ can develop freely.

In the anisotropic case we start from the inversion of the elastic law (4.11), i.e. from the equation

$$\varepsilon_{lm} = C_{ijlm}\sigma^{ij}. \tag{7.25}$$

For a plane stress system it yields the strains

$$\varepsilon_{\gamma\delta} = C_{\alpha\beta\gamma\delta}\,\sigma^{\alpha\beta}, \tag{7.26}$$

$$\varepsilon_{\gamma 3} = C_{\alpha\beta\gamma 3}\,\sigma^{\alpha\beta}, \qquad \varepsilon_{33} = C_{\alpha\beta 33}\,\sigma^{\alpha\beta}.$$

In most cases $C_{\alpha\beta\gamma 3} = 0$, and we will assume that this is true. When (7.26) is inverted, the result has the form

$$\sigma^{\alpha\beta} = \bar{E}^{\alpha\beta\gamma\delta}\varepsilon_{\gamma\delta}, \tag{7.27}$$

and the $\bar{E}^{\alpha\beta\gamma\delta}$ are different from the $E^{ijlm}$ of the three-dimensional law, as explained on page 108. In this case the strain equations (7.26) are a segment of the three-dimensional system, and the stress equations are an inversion of this segment.

In the isotropic case we start from (4.27) and let $\sigma_3^\alpha = \sigma_\alpha^3 = \sigma_3^3 = 0$. Then

$$\left.\begin{aligned} E\varepsilon_\beta^\alpha &= (1+\nu)\sigma_\beta^\alpha - \nu\sigma_\xi^\xi \delta_\beta^\alpha, \\ \varepsilon_3^\alpha = \varepsilon_\alpha^3 = 0, \qquad E\varepsilon_3^3 &= -\nu\sigma_\zeta^\zeta, \end{aligned}\right\} \tag{7.28}$$

and (7.28) can be inverted using the technique demonstrated on (7.8). The result is the following equation:

$$\sigma_\beta^\alpha = \frac{E}{1+\nu}\left(\varepsilon_\beta^\alpha + \frac{\nu}{1-\nu}\varepsilon_\zeta^\zeta \delta_\beta^\alpha\right). \tag{7.29}$$

Once the elastic law has been obtained, the procedure is the same as for plane strain. The differential equation taking the place of (7.12) differs from it only through the appearance of the moduli $\bar{E}^{\alpha\beta\gamma\delta}$ of (7.27) instead of $E^{\alpha\beta\gamma\delta}$. In the isotropic case, (7.14) must be replaced by

$$u^\alpha|_\beta^\beta + \frac{1+\nu}{1-\nu}u^\beta|_\beta^\alpha + \frac{2(1+\nu)}{E}X^\alpha = 0, \tag{7.30}$$

which differs only very little from it.

In the approach using the stress function, the definition (7.15) and its alternate (7.17) remain unchanged, and so does the compatibility equation (7.18). For the anisotropic material, (7.26) replaces (7.7), which means that in (7.20) the compliances $C_{\alpha\beta\gamma\delta}$ of the three-dimensional law replace the compliances $\bar{C}_{\alpha\beta\gamma\delta}$ of plane strain. In the isotropic case (7.9) is to be replaced by (7.28), which ultimately leads to the differential equation

$$\Phi|_{\alpha\beta}^{\alpha\beta} = (1-\nu)\Omega|_\alpha^\alpha, \tag{7.31}$$

which for vanishing body forces reduces to (7.24).

## 7.3. Generalized Plane Strain

We now come back to the more general case of plane strain mentioned briefly on page 108. Figure 7.2 illustrates an anisotropic material of the more general type. The shading indicates the direction of some grain or fibers, i.e. a principal direction of elasticity. When a tensile stress $\sigma^{11}$ is applied, the right angles of the element will not be preserved unless a shear stress $\sigma^{31}$ of a certain magnitude is also applied. Both stresses occur if the deformation consists of nothing but a strain $\varepsilon_{11}$. The shear stresses $\sigma^{13}$ and $\sigma^{23}$, which thus are inevitable, are subjected to an equilibrium condition, which follows from (6.6) with $i = 3$ and $X^3 = 0$:

$$\sigma^{\beta 3}|_\beta = 0, \tag{7.32}$$

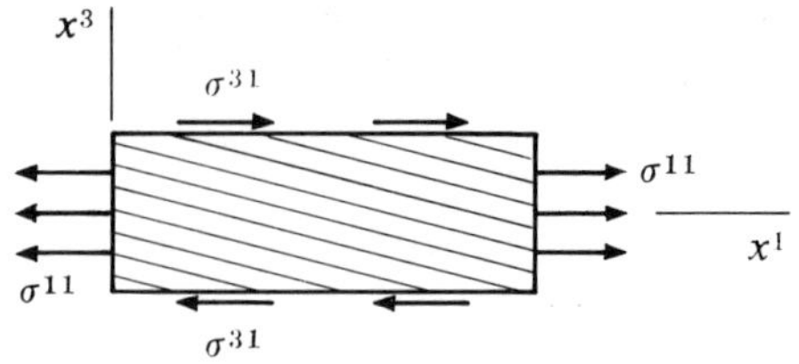


FIGURE 7.2 *Element of an anisotropic material.*

and this equation upsets the balance between the number of unknowns and of equations (the two new stresses are covered by two additional component equations in Hooke's law and the corresponding new strains by kinematic relations). This makes it necessary to admit the existence of a third displacement component $u_3 = w$, which is a function of the $x^\alpha$, but, of course, not of $x^3$, and represents a warping of sections $x^3 = \text{const}$.

We write Hooke's law for this general anisotropic material in the inverted form

$$\varepsilon_{\gamma\delta} = C_{\alpha\beta\gamma\delta}\sigma^{\alpha\beta} + C_{\alpha3\gamma\delta}\sigma^{\alpha3} + C_{3\beta\gamma\delta}\sigma^{3\beta} + C_{33\gamma\delta}\sigma^{33}.$$

The second and third terms represent the same sum and may be combined into one:

$$\varepsilon_{\gamma\delta} = C_{\alpha\beta\gamma\delta}\sigma^{\alpha\beta} + 2C_{\alpha3\gamma\delta}\sigma^{\alpha3} + C_{33\gamma\delta}\sigma^{33} \tag{7.33a}$$

and, similarly, we may write

$$\varepsilon_{\gamma3} = C_{\alpha\beta\gamma3}\sigma^{\alpha\beta} + 2C_{\alpha3\gamma3}\sigma^{\alpha3} + C_{33\gamma3}\sigma^{33}, \tag{7.33b}$$

$$\varepsilon_{33} = C_{\alpha\beta33}\sigma^{\alpha\beta} + 2C_{\alpha333}\sigma^{\alpha3} + C_{3333}\sigma^{33} = 0. \tag{7.33c}$$

The kinematic relation (7.4) is still valid and leads again to the compatibility condition (7.18), but now we have the additional relation

$$\varepsilon_{\gamma3} = \tfrac{1}{2}(u_\gamma|_3 + u_3|_\gamma).$$

Since $u_\gamma$ does not depend upon $x^3$ and since we want to set $u_3 = w$, we write this kinematic relation as

$$\varepsilon_{\gamma3} = \tfrac{1}{2}w|_\gamma. \tag{7.34}$$

By differentiating it with respect to $x^\nu$ we obtain a new compatibility condition:

$$\varepsilon_{\gamma3}|_\nu = \tfrac{1}{2}w|_{\gamma\nu} = \tfrac{1}{2}w|_{\nu\gamma} = \varepsilon_{\nu3}|_\gamma.$$

We write it in the form

$$\varepsilon_{\gamma3}|_\nu \epsilon^{\gamma\nu} = 0. \tag{7.35}$$

Finally, we have the equilibrium conditions. For vanishing body forces they are

$$\sigma^{\alpha\beta}|_{\beta} = 0, \qquad \sigma^{\alpha 3}|_{\alpha} = 0. \tag{7.36a, b}$$

It may be left to the reader to extend the theory to the case of conservative body forces obeying (7.16).

Omitting all duplications caused by the symmetry of the tensors involved, we have in Hooke's law (7.33), the compatibility conditions (7.18) and (7.35), and the equilibrium conditions (7.36), a total of $6 + 2 + 3 = 11$ equations for as many unknowns, namely 6 stresses and 5 strains, not including $\varepsilon_{33}$, which is known to be zero.

One of the equilibrium conditions (7.36) may again be satisfied identically by introducing the stress function $\Phi$ according to (7.15). To take care of (7.36b), we use a second stress function $\Psi$, for which

$$\sigma^{\alpha 3} = \epsilon^{\alpha\lambda}\Psi|_{\lambda}. \tag{7.37}$$

The two stress functions and the stress $\sigma^{33}$ are the surviving unknowns of the problem, and for these we shall now establish three equations. Two of them are based on the compatibility conditions (7.18) and (7.35). We use the elastic law to introduce the stresses and then (7.15) and (7.37) to introduce $\Phi$ and $\Psi$ and thus obtain the following differential equations, valid for any anisotropic, yet homogeneous material:

$$(C_{\alpha\beta\gamma\delta}\Phi|_{\lambda\mu\nu\rho}\,\epsilon^{\alpha\lambda}\epsilon^{\beta\mu} + 2C_{\alpha 3\gamma\delta}\Psi|_{\lambda\nu\rho}\,\epsilon^{\alpha\lambda} + C_{33\gamma\delta}\sigma^{33}|_{\nu\rho})\epsilon^{\gamma\nu}\epsilon^{\delta\rho} = 0, \tag{7.38a}$$

$$(C_{\alpha\beta\gamma 3}\Phi|_{\lambda\mu\nu}\,\epsilon^{\alpha\lambda}\epsilon^{\beta\mu} + 2C_{\alpha 3\gamma 3}\Psi|_{\lambda\nu}\,\epsilon^{\alpha\lambda} + C_{33\gamma 3}\sigma^{33}|_{\nu})\epsilon^{\gamma\nu} = 0. \tag{7.38b}$$

The third equation of the set is simply (7.33c) written in terms of the stress functions:

$$C_{\alpha\beta 33}\Phi|_{\lambda\mu}\,\epsilon^{\alpha\lambda}\epsilon^{\beta\mu} + 2C_{\alpha 333}\Psi|_{\lambda}\,\epsilon^{\alpha\lambda} + C_{3333}\sigma^{33} = 0. \tag{7.38c}$$

Since this is an algebraic equation, it may easily be used to eliminate $\sigma^{33}$ from the other two with the following result:

$$[(C_{\alpha\beta\gamma\delta}C_{3333} - C_{\alpha\beta 33}C_{33\gamma\delta})\Phi|_{\lambda\mu\nu\rho}\,\epsilon^{\beta\mu} + 2(C_{\alpha 3\gamma\delta}C_{3333} - C_{\alpha 333}C_{33\gamma\delta})\Psi|_{\lambda\nu\rho}]\epsilon^{\alpha\lambda}\epsilon^{\gamma\nu}\epsilon^{\delta\rho} = 0, \tag{7.39a}$$

$$[(C_{\alpha\beta\gamma 3}C_{3333} - C_{\alpha\beta 33}C_{33\gamma 3})\Phi|_{\lambda\mu\nu}\,\epsilon^{\beta\mu} + 2(C_{\alpha 3\gamma 3}C_{3333} - C_{\alpha 333}C_{33\gamma 3})\Psi|_{\lambda\nu}]\epsilon^{\alpha\lambda}\epsilon^{\gamma\nu} = 0. \tag{7.39b}$$

This is a system of two simultaneous differential equations for $\Phi$ and $\Psi$. It is of the sixth order, while we found one fourth-order equation when the material was isotropic or had plane anisotropy.

It is interesting to specialize (7.39) for plane anisotropy. It amounts to setting

$$C_{\alpha 3\gamma\delta} = C_{\alpha 333} = C_{\alpha\beta\gamma 3} = C_{33\gamma 3} = 0.$$

This uncouples the equations, which now read

$$(C_{\alpha\beta\gamma\delta} C_{3333} - C_{\alpha\beta 33} C_{33\gamma\delta})\Phi|_{\lambda\mu\nu\rho}\, \epsilon^{\alpha\lambda}\epsilon^{\beta\mu}\epsilon^{\gamma\nu}\epsilon^{\delta\rho} = 0, \tag{7.40a}$$

$$C_{\alpha 3\gamma 3}\, \Psi|_{\lambda\nu}\, \epsilon^{\alpha\lambda}\epsilon^{\gamma\nu} = 0. \tag{7.40b}$$

In addition there is (7.38c) for $\sigma^{33}$, which reduces to

$$C_{3333}\,\sigma^{33} = -C_{\alpha\beta 33}\,\Phi|_{\lambda\mu}\,\epsilon^{\alpha\lambda}\epsilon^{\beta\mu}. \tag{7.40c}$$

Equations (7.40a, c) are connected with the stress system $\sigma^{\alpha\beta}$, $\sigma^{33}$, which we have studied before, and we surmise that (7.40a) is identical with the homogeneous form ($\Omega \equiv 0$) of (7.20). This may easily be verified. The elastic law consists in this case of the two equations

$$\varepsilon_{\gamma\delta} = C_{\alpha\beta\gamma\delta}\,\sigma^{\alpha\beta} + C_{3\,3\gamma\delta}\,\sigma^{33},$$
$$\varepsilon_{33} = C_{\alpha\beta 33}\,\sigma^{\alpha\beta} + C_{3333}\,\sigma^{33} = 0$$

and after elimination of $\sigma^{33}$ this yields the relation

$$\varepsilon_{\gamma\delta} = \frac{C_{\alpha\beta\gamma\delta}\,C_{3333} - C_{\alpha\beta 33}\,C_{3\,3\gamma\delta}}{C_{3333}}\,\sigma^{\alpha\beta}.$$

Comparison with (7.7) shows that the coefficient between parentheses in (7.40a) equals $\bar{C}_{\alpha\beta\gamma\delta}\,C_{3333}$, and the scalar $C_{3333}$ may, of course, be factored out and dropped.

Equation (7.40b) represents a second problem, which we have not treated as part of the plane strain problem. The stress function $\Psi$ is related to the shear stresses $\sigma^{\alpha 3}$ normal to planes $x^3 = \text{const}$, and the compliance $C_{\alpha 3\gamma 3}$ is a shear compliance connecting these stresses with the strains $\varepsilon_{\gamma 3}$, as may be seen from (7.33b), which for plane anisotropy simplifies to read

$$\varepsilon_{\gamma 3} = 2C_{\alpha 3\gamma 3}\,\sigma^{\alpha 3} = C_{\alpha 3\gamma 3}\,\sigma^{\alpha 3} + C_{3\alpha\gamma 3}\,\sigma^{3\alpha}.$$

This shear problem is of the second order and is a special case of Milne-Thomson's antiplane problem [23]. In the case of general anisotropy it is coupled with the ordinary plane strain problem.

## 7.4. Torsion

We consider a straight bar of constant cross section (a prism or a cylinder) and choose two coordinates $x^\alpha$ in the plane of a cross section and a third coordinate $x^3 = z$ measured in the direction of the generators of the cylindrical

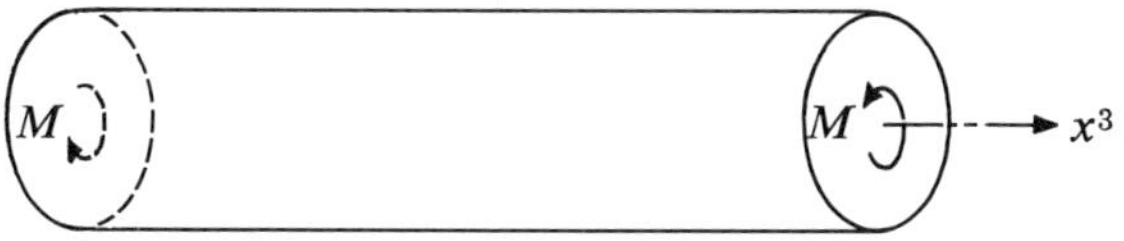


FIGURE 7.3 *Torsion bar.*

surface. This is again a monoclinic coordinate system (see p. 106) with $g_{\alpha 3} = 0$, $g_{33} = 1$.

The bar is loaded by couples $M$ at both ends (Figure 7.3). In every cross section, stresses must act which transmit this *torque* from one side to the other. They are shear stresses $\sigma^{3\alpha} = \sigma_3^\alpha$.

The deformation consists in the first place of the rotation of the cross sections, each in its plane, about an axis parallel to the $z$ axis. Since every element of the bar of length $dz$ undergoes the same deformation, the angle of rotation $\psi$ is a linear function of $z$, in the simplest case proportional to $z$:

$$\psi = \theta z. \tag{7.41}$$

The quantity $\theta$ is called the twist of the bar.

During the rotation the cross section does not change its shape, that is, the in-plane strain components vanish:

$$\varepsilon_{\alpha\beta} = 0, \qquad \varepsilon_\beta^\alpha = 0. \tag{7.42}$$

From (6.1) it follows that

$$u_\alpha|_\beta + u_\beta|_\alpha = 0 \tag{7.43}$$

and hence

$$u_\alpha|_{\beta\gamma} = -u_\beta|_{\alpha\gamma}.$$

Since the sequence of covariant differentiations is interchangeable, we conclude from (7.43) that

$$u_\alpha|_{\beta\gamma} = u_\alpha|_{\gamma\beta} = -u_\gamma|_{\alpha\beta} = -u_\gamma|_{\beta\alpha} = +u_\beta|_{\gamma\alpha} = u_\beta|_{\alpha\gamma}$$

and this contradicts the preceding equation, unless

$$u_\alpha|_{\beta\gamma} = 0.$$

We conclude that this statement must be true.

Because of (7.42), each area element of the cross section rotates in its plane like a rigid body and the component $\omega^3$ of the "average" rotation defined by (6.3) is simply the rotation $\psi$ of the cross section:

$$\omega^3 = -\tfrac{1}{2} u_\alpha|_\beta \, \epsilon^{\alpha\beta} = \psi = \theta z. \tag{7.44}$$

In addition to the rigid-body rotation of each cross section in its own plane, a warping of the section takes place, consisting of a displacement $u^3 = w$ normal to the plane of the cross section, dependent on $x^\alpha$, but independent of $z$. This leads to the strains

$$\varepsilon_{33} = \varepsilon_3^3 = w|_3 = 0, \tag{7.45a}$$

$$\varepsilon_{\alpha 3} = \tfrac{1}{2}(u_\alpha|_3 + u_3|_\alpha) = \tfrac{1}{2}(u_\alpha|_3 + w|_\alpha), \tag{7.45b}$$

$$\varepsilon_3^\alpha = \varepsilon_{\beta 3}\, g^{\beta\alpha} + \varepsilon_{33}\, g^{3\alpha} = \tfrac{1}{2}(u^\alpha|_3 + w|^\alpha). \tag{7.45c}$$

Since $w|_{\alpha\beta} = w|_{\beta\alpha}$ we have $w|_{\alpha\beta}\,\epsilon^{\alpha\beta} = 0$ and find from (7.44) and (7.45b) that

$$\varepsilon_{3\alpha}|_\beta\, \epsilon^{\alpha\beta} = \tfrac{1}{2}(u_\alpha|_{3\beta} + w|_{\alpha\beta})\epsilon^{\alpha\beta} = \tfrac{1}{2}u_\alpha|_{\beta 3}\, \epsilon^{\alpha\beta} = -\theta \tag{7.46}$$

and also

$$\varepsilon_3^\alpha|^\beta \epsilon_{\alpha\beta} = -\theta. \tag{7.47}$$

This completes the kinematics of the deformation.

When we introduce the strains from (7.42) and (7.45a) in Hooke's law (4.25), we see that

$$\sigma_\beta^\alpha = \frac{E}{1+\nu}\left[\varepsilon_\beta^\alpha + \frac{\nu}{1-2\nu}(\varepsilon_\lambda^\lambda + \varepsilon_3^3)\delta_\beta^\alpha\right] = 0$$

and also $\sigma_3^3$ turns out to be zero. The only nonvanishing stress components are $\sigma_3^\alpha = \sigma^{3\alpha}$ in the cross section and $\sigma_3^\alpha = \sigma^{\alpha 3}$ in longitudinal sections. The former ones transmit the torque $M$ and form a plane vector field $\sigma_3^\alpha \mathbf{g}_\alpha$, for which we may use the alternate notation $\tau^\alpha \mathbf{g}_\alpha$. From (4.25) we have

$$\tau^\alpha = \sigma_3^\alpha = \frac{E}{1+\nu}\,\varepsilon_3^\alpha. \tag{7.48}$$

In combination with (7.47) this leads to

$$\tau^\alpha|^\beta \epsilon_{\alpha\beta} = \frac{E}{1+\nu}\,\varepsilon_3^\alpha|^\beta \epsilon_{\alpha\beta} = -\frac{E\theta}{1+\nu}. \tag{7.49}$$

The last item of physical information needed is the equilibrium condition (6.6). We drop the body force term and let $i = 3$, $j = \alpha$:

$$\sigma^{\alpha 3}|_\alpha \equiv \tau^\alpha|_\alpha = 0. \tag{7.50}$$

This equation is identically satisfied if we write the stress components as the derivatives of a stress function $\Phi$:

$$\tau^\alpha = \Phi|_\gamma\, \epsilon^{\gamma\alpha}. \tag{7.51}$$

When this is introduced in (7.49), a differential equation for $\Phi$ results:

$$\Phi|_{\gamma}^{\beta}\,\epsilon_{\alpha\beta}\,\epsilon^{\gamma\alpha} = -\frac{E\theta}{1+\nu}.$$

With the help of (3.29b) the left-hand side can be simplified to read

$$\Phi|_{\gamma}^{\beta}\,\epsilon_{\alpha\beta}\,\epsilon^{\gamma\alpha} = -\Phi|_{\gamma}^{\beta}\,\epsilon_{\alpha\beta}\,\epsilon^{\alpha\gamma} = -\Phi|_{\gamma}^{\beta}\,\delta_{\beta}^{\gamma} = -\Phi|_{\beta}^{\beta}$$

and this ultimately leads to Poisson's equation

$$\Phi|_{\alpha}^{\alpha} = \frac{E\theta}{1+\nu} = 2G\theta. \tag{7.52}$$

This is the differential equation of the torsion problem. It should be noted that in deriving it we started from the assumption expressed by equations (7.42), but that in the course of the derivation it turned out that all the equations of the theory of elasticity could be satisfied. Therefore, the initial assumption does not have the character of an approximation made to simplify the analysis, but rather that of a partial anticipation of the result of an exact theory.

In most problems of elasticity, the boundary conditions are obvious since they consist of prescribing a number of mechanical quantities (stresses or displacements) along the boundary. In the present problem the boundary condition to be imposed upon $\Phi$ needs closer inspection.

The only load to be applied to the torsion bar is the torque $M$ at its ends. The cylindrical surface is to be free of external tractions, in particular of shear stresses $\sigma^{\alpha 3}$ and, hence, the shear stress vector $\tau^{\alpha}$ at boundary points of the cross section must not have a component normal to the boundary. Let $dx^{\beta}\,\mathbf{g}_{\beta}$ be a line element vector of the boundary; then the fact that the stress vector $\tau^{\alpha}\mathbf{g}_{\alpha}$ is parallel to it may be expressed in the form that the cross product of both vectors vanishes:

$$\tau^{\alpha}\,dx^{\beta}\,\epsilon_{\alpha\beta} = 0.$$

Making use of (7.51) and (3.29b), we bring this into the form

$$\Phi|_{\gamma}\,\epsilon^{\gamma\alpha}\,dx^{\beta}\,\epsilon_{\alpha\beta} = -\Phi|_{\gamma}\,\delta_{\beta}^{\gamma}\,dx^{\beta} = -\Phi|_{\beta}\,dx^{\beta} = 0.$$

From this we conclude that along the boundary

$$\Phi = \text{const.} \tag{7.53}$$

If the cross section is simply connected, we may set on its boundary $\Phi = 0$, because any other choice of the boundary value would only add a constant to $\Phi(x^{\gamma})$ in the entire field, and we are only interested in the derivative $\Phi|_{\gamma} = \Phi_{,\gamma}$, in which an additive constant does not show up. If the cross section has one or several holes, its boundary consists of several separate curves. We still may assign the value $\Phi = 0$ to the exterior boundary, but the values of $\Phi$ along the edges of the holes are not free for arbitrary choice, but follow from the requirement that the warping displacement $w$ is a unique function of $x^\alpha$. To formulate this requirement in terms of $\tau^\gamma$ or of $\Phi$, we start from (7.45c) and (7.48):

$$\tau_\alpha = 2G\varepsilon_{\alpha 3} = G(u_\alpha|_3 + w|_\alpha).$$

We solve for $w|_\alpha = w_{,\alpha}$ and integrate along a closed curve which lies entirely inside the material cross section, that is, in the domain in which $w$ is defined. Uniqueness of $w$ requires that this integral vanishes:

$$\oint w|_\alpha \, dx^\alpha = \frac{1}{G}\oint \tau_\alpha \, dx^\alpha - \oint u_\alpha|_3 \, dx^\alpha = 0. \tag{7.54}$$

The plane vector $u_\alpha$ describes the rigid-body rotation of the cross section in its own plane. Although it has physical significance only in the material cross section, it is uniquely defined in the entire plane including the holes. Therefore, we may apply Stokes' theorem (5.42) to the last integral in (7.54):

$$\oint u_\alpha|_3 \, dx^\alpha = \tfrac{1}{2}\epsilon^{\beta\alpha}\epsilon_{\lambda\mu}\int_B u_\alpha|_{3\beta} \, dr^\lambda \, dt^\mu = \epsilon^{\beta\alpha}\int_B u_\alpha|_{\beta 3} \, dB$$

in which $B$ is the area enclosed by the integration path and

$$\tfrac{1}{2}\epsilon_{\lambda\mu} \, dr^\lambda \, dt^\mu = dB$$

is an area element. We use (7.46) to express $u_\alpha$ in terms of $\theta$:

$$\oint u_\alpha|_3 \, dx^\alpha = 2\int_B \theta \, dB = 2\theta B$$

and equate this to the stress integral in (7.54), which may be written in terms of $\Phi$. This ultimately yields the condition

$$\oint \Phi|^\gamma \epsilon_{\gamma\alpha} \, dx^\alpha = 2G\theta B, \tag{7.55}$$

which the stress function must satisfy when the integral is extended over any one of the interior boundaries of the cross section (the edges of the holes).

For a cross section with $n$ holes the practical computation is done in the following way: One first finds the solution $\Phi_{(0)}$ of the differential equation (7.52) which satisfies the condition $\Phi = 0$ on *all* boundaries. Then one calculates solutions $\Phi_{(k)}$, $k = 1, 2, \ldots, n$ of the homogeneous equation $\Phi|^\alpha_\alpha = 0$, prescribing $\Phi = 1$ along the boundary of the $k$th hole and $\Phi = 0$ on all other boundaries. For each of the solutions one evaluates the integral

(7.55) for each of the interior boundaries. The correct solution is then found by linear superposition:

$$\Phi = \Phi_{(0)} + \sum_{k=1}^{n} C_{(k)} \Phi_{(k)},$$

the free constants $C_{(k)}$ being determined from $n$ linear equations resulting from the application of the boundary condition (7.55) to each of the interior boundaries. The subscripts in parentheses are not subject to the range convention and the summation convention and do not imply tensor character.

Thus far, we have solved the torsion problem in terms of the twist $\theta$. Usually it is not the twist that is known in advance, but rather the torque $M$. To make our solution useful, we must find a relation between the two quantities. In addition, we might want to assure ourselves that the stress field $\tau^\alpha$ in the cross section can really be reduced to a couple and does not have a resultant force.

Here we stand before an intrinsic difficulty of the tensorial method. The resultant force (if there should be one) is a vector which is not attached to a specific point in the plane of the cross section; but the base vectors $\mathbf{g}_\alpha$, which we need for defining vector components, vary from point to point and, therefore, there is no reference frame to which the resultant might be referred. A similar difficulty arises for the torque. In principle, it is a vector, but its only component points in the direction $\mathbf{g}_3$. Therefore, with respect to the frame $\mathbf{g}_\alpha$ it is a scalar. But to write the moment, we need the lever arm of each infinitesimal force $\tau^\alpha \, dA$ with respect to some fixpoint, and this again is a finite vector not attached to the metric of any particular point of the plane. We shall see how, in the present case, this difficulty can be avoided.

We can deal rather easily with the question whether or not the shear stresses have a resultant $\mathbf{R}$. If there is one, it will have to be the same in all cross sections, and when we cut a piece of arbitrary length $l$ from the bar, there will be two forces acting on its ends as shown in Figure 7.4. They form a couple, which requires some other forces for equilibrium. Since $\tau^\alpha = \sigma^{3\alpha}$ and $\sigma^{\alpha 3}$ are the only nonvanishing stress components and since $\sigma^{\alpha 3} = 0$ on the surface of the bar, we know that there are no forces with which the couple

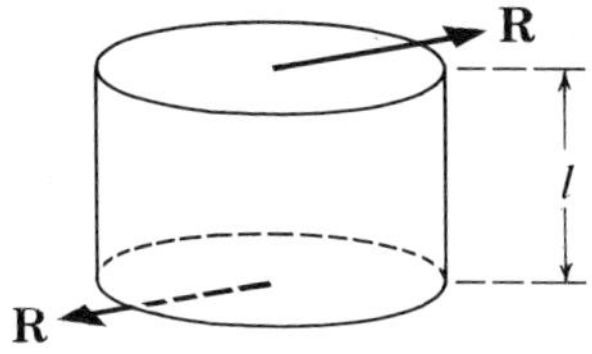


FIGURE 7.4 *Finite length element of a torsion bar.*

$l\mathbf{R}$ could be in equilibrium. On the other hand, the stress system satisfies the equilibrium condition (6.6), which assures the equilibrium of each volume element and, hence, also that of the entire piece shown in Figure 7.4. Thus we are forced to admit that $\mathbf{R} = 0$.

We now turn our attention to the torque $M$. Since the shear stresses $\tau^\alpha$ have no resultant, $M$ is their moment with respect to any reference point $O$ in the plane of the cross section. In (7.51) we have connected $\tau^\alpha$ with a scalar function $\Phi$. We may plot the values of $\Phi$ as ordinates normal to the (horizontal) $x^\alpha$ plane. This yields a surface, which covers the area of the cross section, rising from zero values at the boundary. It is known as the *stress hill.* At any point of the section the gradient vector $\Phi|_\beta \mathbf{g}^\beta$ points in the direction of the steepest ascent of the hill and, according to a statement made on page 36, the shear stress is at right angles to this direction, hence tangential to a line $\Phi = \text{const}$.

We cut the hill by two horizontal planes at levels $\Phi$ and $\Phi + d\Phi$ and project the intersection curves into the $x^\alpha$ plane. There they form two concentric loops enclosing between them a narrow strip as shown in Figure 7.5. We consider an area element of this strip, described by the vectors $d\mathbf{s}$ and $d\mathbf{t}$. Its area is

$$d\mathbf{t} \times d\mathbf{s} = dt^\delta\, ds^\beta\, \epsilon_{\delta\beta}\, \mathbf{g}^3 = dA_3\; \mathbf{g}^3 = dA\; \mathbf{g}^3.$$

The shear force acting in this element is

$$\boldsymbol{\tau} dA = \Phi|_\gamma\, \epsilon^{\gamma\alpha} \mathbf{g}_\alpha\, ds^\beta\, dt^\delta\, \epsilon_{\delta\beta}\,.$$

Applying (3.29a), we write it in the form

$$\tau^\alpha\, dA = \Phi|_\gamma\, ds^\beta\, dt^\delta (\delta^\gamma_\delta \delta^\alpha_\beta - \delta^\gamma_\beta \delta^\alpha_\delta) = \Phi|_\gamma (ds^\alpha\; dt^\gamma - ds^\gamma\; dt^\alpha).$$

According to (5.21), $\Phi|_\gamma\, ds^\gamma$ is the change of $\Phi$ along the vector $d\mathbf{s}$, which is zero since the vector follows a line $\Phi = \text{const}$, and $\Phi|_\gamma\, dt^\gamma = d\Phi$, the change of $\Phi$ across the strip. Therefore,

$$\tau^\alpha\, dA = d\Phi\, ds^\alpha,$$

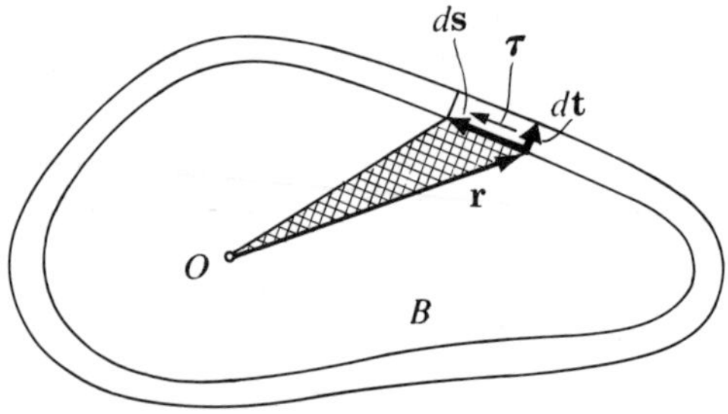


FIGURE 7.5 *Torsion bar, lines* $\Phi = \text{const}$ *in the cross section.*

which confirms that the shear vector has the direction of the lines $\Phi = \text{const.}$

The moment of the force $\tau^\alpha\, dA\, \mathbf{g}_\alpha$ with respect to the point $O$ is

$$\mathbf{r} \times \tau^\alpha\, dA\, \mathbf{g}_\alpha = \mathbf{r} \times d\mathbf{s}\, d\Phi$$

and this is twice the product of $d\Phi$, the triangle shaded in Figure 7.5, and the unit vector $\mathbf{g}_3$. The sum of the moments of all the shear stresses acting in the strip is $2B\, d\Phi$, where $B$ is the area enclosed by any of the curves bounding the strip. The product $2B\, d\Phi$ is twice the volume of a horizontal slice of the stress hill of thickness $d\Phi$. It is a contribution $dM$ to the torque $M$. The entire torque is twice the sum of all such slices, that is, twice the volume of the stress hill. This may be expressed by the integral of products of $\Phi$ and area elements $dA$ of the cross section:

$$M = 2 \int \Phi\, dA = 2 \iint \Phi\, ds^\beta\, dt^\alpha \epsilon_{\alpha\beta}\,. \tag{7.56}$$

It is now possible to solve the differential equation (7.52) of the torsion problem for an assumed value of the twist $\theta$, then calculate the torque $M$ from (7.56) and adapt the solution to any prescribed torque value by simply applying a proportionality factor.

## 7.5. Plates

We again consider a thin plane sheet of elastic material, but this time we assume that it undergoes a deflection normal to its middle plane. We again use a coordinate system consisting of two coordinates $x^\alpha$ in the middle plane and a third coordinate $x^3$ normal to it. We choose as base vector $\mathbf{g}_3$ a unit vector, and then $x^3 = z$ is simply the distance of a point from the middle plane. Let us assume that the middle plane is horizontal and that $\mathbf{g}_3$ points upward.

Figure 7.6 shows a section through the plate before and after deformation. The $x^1$ axis lies in the middle plane and $AB$ is a normal to it. When a load is

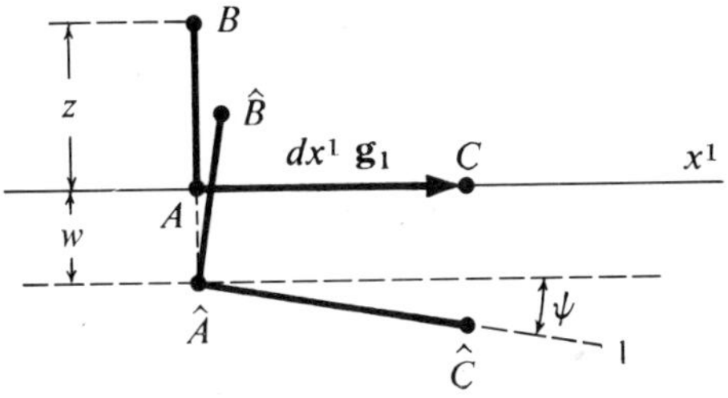


FIGURE 7.6 *Section through a plate before and after deformation.*

applied, the point $A$ moves downward by the amount $w$ to position $\hat{A}$, hence $w = -u_3$.

In the deformed plate, line elements $AC$ of the middle plane are no longer horizontal and normals like $AB$ are no longer vertical. The difference between the rotations of both is a shear deformation of the plate material. In thin plates, as in slender beams, this shear strain is small enough to be neglected, and plate theory is built on the assumption of the *conservation of normals*: All material points that lie on a normal to the middle plane before deformation, lie on a normal to the deformed middle plane after deformation.

The shear strain between a normal and the middle plane has two components

$$\varepsilon_{\alpha 3} = \tfrac{1}{2}(u_\alpha|_3 + u_3|_\alpha).$$

When this equals zero, then $u_\alpha|_3 = -u_3|_\alpha$. Making use of (7.3), we have

$$u_{\alpha,3} = -u_3|_\alpha = w|_\alpha$$

and after integration

$$u_\alpha = w|_\alpha x^3 = w|_\alpha z. \tag{7.57}$$

From the displacement $u_\alpha$ we find the strains parallel to the middle plane

$$\varepsilon_{\alpha\beta} = \tfrac{1}{2}(u_\alpha|_\beta + u_\beta|_\alpha) = \tfrac{1}{2}(w|_{\alpha\beta} + w|_{\beta\alpha})z = w|_{\alpha\beta}\, z. \tag{7.58}$$

Since these strains are proportional to $z$, the stresses to be calculated from Hooke's law also have the same distribution, and in each section of the plate they can be combined to form a couple. We shall now proceed to find these internal moments.

We cut the plate normal to its middle plane, along an arbitrary line element

$$d\mathbf{r} = dx^\gamma\, \mathbf{g}_\gamma .$$

In the area element of height $h$ we define a subelement of height $dz$ as shown in Figure 7.7. Its area is represented by the vector

$$d\mathbf{A} = d\mathbf{r} \times dz\, \mathbf{g}_3 = dx^\gamma\, dz\, \epsilon_{\gamma 3\alpha}\, \mathbf{g}^\alpha = \epsilon_{\alpha\gamma}\, dx^\gamma\, dz\, \mathbf{g}^\alpha,$$

which has the direction of the outer normal if we assume that the material of the plate lies behind the section. In this subelement, the force

$$d\mathbf{F} = dF^\beta\, \mathbf{g}_\beta + dF^3\, \mathbf{g}_3$$

is transmitted. Using the definition (4.6) of stress, we write its components as

$$dF^\beta = \sigma^{\alpha\beta} \epsilon_{\alpha\gamma}\, dx^\gamma\, dz,$$
$$dF^3 = \sigma^{\alpha 3} \epsilon_{\alpha\gamma}\, dx^\gamma\, dz.$$

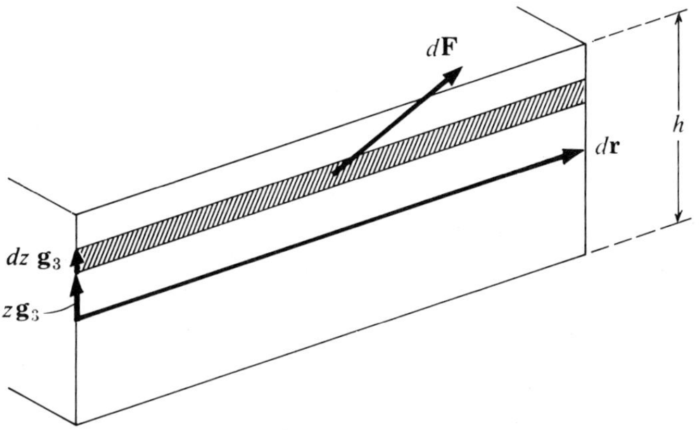


FIGURE 7.7 *Section through a plate.*

The vertical components $dF^3\mathbf{g}_3$ in all the subelements have a resultant, the shear force of the plate:

$$\int_{-h/2}^{+h/2} dF^3\,\mathbf{g}_3 = \epsilon_{\alpha\gamma}\,dx^\gamma\,\mathbf{g}_3\int_{-h/2}^{+h/2}\sigma^{\alpha 3}\,dz = -\epsilon_{\alpha\gamma}\,dx^\gamma\,Q^\alpha\mathbf{g}_3\,. \tag{7.59}$$

In the same way, we may write the resultant of the in-plane components $dF^\beta\mathbf{g}_\beta$:

$$\int_{-h/2}^{+h/2} dF^\beta\,\mathbf{g}_\beta = \epsilon_{\alpha\gamma}\,dx^\gamma\,\mathbf{g}_\beta\int_{-h/2}^{+h/2}\sigma^{\alpha\beta}\,dz = \epsilon_{\alpha\gamma}\,dx^\gamma\,N^{\alpha\beta}\mathbf{g}_\beta\,, \tag{7.60}$$

but we shall soon see that it vanishes in the case of plate bending.

The moment of $d\mathbf{F}$ with respect to the center of the section is

$$z\mathbf{g}_3\times d\mathbf{F} = z\mathbf{g}_3\times(dF^\beta\,\mathbf{g}_\beta + dF^3\,\mathbf{g}_3) = dF^\beta\,z\epsilon_{3\beta\delta}\mathbf{g}^\delta$$

and by integrating across the thickness of the plate we find the resultant moment

$$\int_{-h/2}^{+h/2} dF^\beta\,z\epsilon_{\beta\delta}\mathbf{g}^\delta = \epsilon_{\alpha\gamma}\,\epsilon_{\beta\delta}\,dx^\gamma\,\mathbf{g}^\delta\int_{-h/2}^{+h/2}\sigma^{\alpha\beta}z\,dz = -\epsilon_{\alpha\gamma}\,\epsilon_{\beta\delta}\,dx^\gamma\,M^{\alpha\beta}\mathbf{g}^\delta. \tag{7.61}$$

From (7.59) through (7.61) we extract the following definitions for the (transverse) shear force

$$Q^\alpha = -\int_{-h/2}^{+h/2}\sigma^{\alpha 3}\,dz, \tag{7.62a}$$

the tension-and-shear tensor

$$N^{\alpha\beta} = \int_{-h/2}^{+h/2}\sigma^{\alpha\beta}\,dz, \tag{7.62b}$$

and the moment tensor

$$M^{\alpha\beta} = -\int_{-h/2}^{+h/2} \sigma^{\alpha\beta} z \, dz. \tag{7.62c}$$

These quantities are called *stress resultants*, and they take the place of the stresses when we describe the stress system of the plate. The minus signs in two of the definitions have been added arbitrarily. They reflect the sign conventions traditionally used in plate theory.

We shall now try to visualize the meaning of these definitions by applying them to sections along coordinate lines of a skew, rectilinear coordinate system. Figure 7.8 shows a plate element cut out along such lines. For the line element vector $AB$ the component $dx^1 = 0$ and the last member of (7.61) reduces to

$$-\epsilon_{12}\,\epsilon_{\beta\delta}\,dx^2\,M^{1\beta}\mathbf{g}^\delta = -\epsilon_{12}\,dx^2(\epsilon_{12}M^{11}\mathbf{g}^2 + \epsilon_{21}M^{12}\mathbf{g}^1).$$

The two components of this moment vector are shown in the figure in the directions corresponding to positive values of $M^{11}$ and $M^{12}$. In rectangular coordinates, $M^{11}$ is the bending moment and $M^{12}$ is the twisting moment, as shown in Figure 7.9 in the commonly used notation.

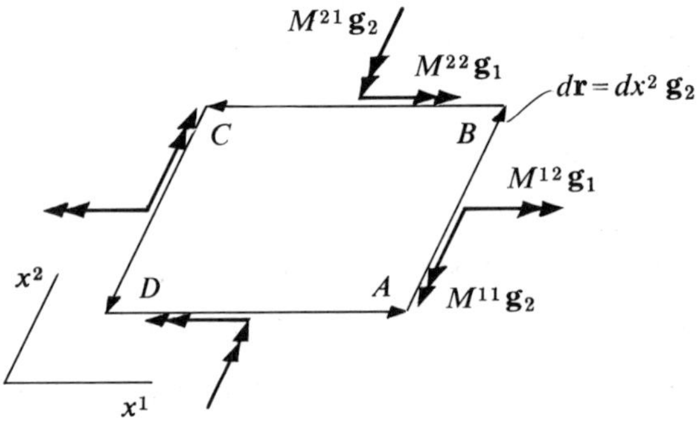


FIGURE 7.8 *Moments acting on a plate element.*

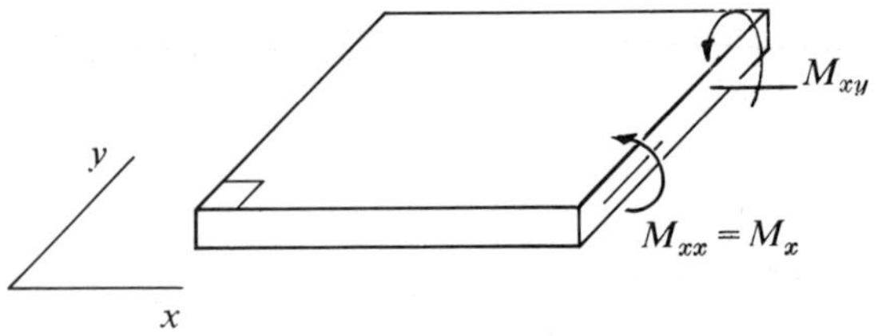


FIGURE 7.9 *Bending and twisting moments acting on a cartesian plate element.*

By repeating the argument for the sides *BC*, *CD*, and *DA*, the reader will find the other moment vectors shown in Figure 7.8. It should be noted that the direction of the line element vectors has been so chosen that the plate element lies to their left. This assures that the vectors $d\mathbf{A}$ of all section elements point in the direction of an outer normal.

The function of the two $\epsilon$ factors in (7.61) goes beyond regulating the positive directions of the moment vectors in Figure 7.8. Because of the factor $\epsilon_{\alpha\gamma}$ the first index in $M^{\alpha\beta}$ is opposed to the one indicating the direction of the section element, and $\epsilon_{\beta\delta}$ makes the second index opposed to that of the base vector $\mathbf{g}^\delta$.

The shear force components $Q^\alpha$ have only one index and, therefore, are the components of a vector

$$\mathbf{Q} = Q^\alpha \mathbf{g}_\alpha$$

lying in the middle plane of the plate. This vector is not in any sense the resultant of the shear forces $Q^1$ and $Q^2$ acting in two different sections $x^1 = \text{const}$ and $x^2 = \text{const}$ of the plate. The proper meaning of this vector will become clear in a discussion of the principal axes problem on page 180.

Since the stress component $\sigma^{33}$ is either zero or rather small and since the shear stresses $\sigma^{\alpha 3}$ do not affect the in-plane strains $\varepsilon_{\alpha\beta}$, every thin layer of thickness $dz$ in the plate is in a state of plane stress. We may apply the elastic law (7.27) with the meaning of the moduli as defined on page 112, and introduce it into the definitions (7.62b, c). Making use of the kinematic relation (7.58), we find

$$N^{\alpha\beta} = \int_{-h/2}^{+h/2} \bar{E}^{\alpha\beta\gamma\delta} w|_{\gamma\delta}\, z\, dz = \bar{E}^{\alpha\beta\gamma\delta} w|_{\gamma\delta} \int_{-h/2}^{+h/2} z\, dz = 0\,,$$

$$M^{\alpha\beta} = -\bar{E}^{\alpha\beta\gamma\delta} w|_{\gamma\delta} \int_{-h/2}^{+h/2} z^2\, dz = -\bar{E}^{\alpha\beta\gamma\delta} \frac{h^3}{12} w|_{\gamma\delta}\,. \tag{7.63}$$

The first of these equations eliminates $N^{\alpha\beta}$ from further consideration, and (7.63) is what might be called the elastic law of the plate. It should be noted that in this equation $z^2$ and $h^3$ designate powers, not components, of these quantities.

For isotropic plates, we use the elastic law (7.29), which we write in the form

$$\sigma^{\alpha\beta} = \sigma^\beta_\gamma g^{\alpha\gamma} = \frac{E}{1+\nu}\left(\varepsilon_{\gamma\delta}\, g^{\alpha\gamma} g^{\beta\delta} + \frac{\nu}{1-\nu}\, \varepsilon_{\eta\zeta}\, g^{\alpha\beta} g^{\eta\zeta}\right). \tag{7.64}$$

This leads to

$$M^{\alpha\beta} = -\frac{Eh^3}{12(1+\nu)}\, w|_{\gamma\delta}\left(g^{\alpha\gamma} g^{\beta\delta} + \frac{\nu}{1-\nu}\, g^{\alpha\beta} g^{\gamma\delta}\right) = -K[(1-\nu) w|^{\alpha\beta} + \nu w|^\gamma_\gamma\, g^{\alpha\beta}] \tag{7.65}$$

as the elastic law of the plate. The quantity

$$K = \frac{Eh^3}{12(1-\nu^2)} \tag{7.66}$$

is called the bending stiffness of the plate.

There is, of course, no relation of a similar kind for $Q^\alpha$, since the assumption of the conservation of normals denies the existence of a deformation directly attributable to the shear force.

The equilibrium of a plate element like the one shown in Figure 7.8 can be expressed in terms of the stress resultants. To obtain the equilibrium conditions, we choose a formal procedure, in which the plate element will never appear. We start from the general equilibrium condition (6.6) and rewrite it here for $i = \alpha$. Splitting the summation over $j = 1, 2, 3$ into a summation over $\beta = 1, 2$ and an extra term for $j = 3$, we have

$$\sigma^{\alpha\beta}|_\beta + \sigma^{\alpha 3}|_3 + X^\alpha = 0.$$

We multiply this equation by $z\,dz$ and integrate:

$$\int_{-h/2}^{+h/2} \sigma^{\alpha\beta}|_\beta\, z\,dz + \int_{-h/2}^{+h/2} \sigma^{\alpha 3}|_3\, z\,dz + \int_{-h/2}^{+h/2} X^\alpha z\,dz = 0. \tag{7.67}$$

In the first term we recognize the covariant derivative of the moment $M^{\alpha\beta}$, and in the second term we have

$$\sigma^{\alpha 3}|_3 = \sigma^{\alpha 3}{}_{,3} + \sigma^{\gamma 3}\Gamma^\alpha_{\gamma 3} + \sigma^{\alpha\gamma}\Gamma^3_{\gamma 3} = \sigma^{\alpha 3}{}_{,3}$$

because of (7.1) and the statement attached to these equations. This allows integration by parts:

$$\int_{-h/2}^{+h/2} \sigma^{\alpha 3}{}_{,3}\, z\,dz = \Big[\sigma^{\alpha 3} z\Big]_{-h/2}^{+h/2} - \int_{-h/2}^{+h/2} \sigma^{\alpha 3}\,dz = Q^\alpha + \Big[\sigma^{\alpha 3} z\Big]_{-h/2}^{+h/2}$$

and this brings (7.67) into the form

$$-M^{\alpha\beta}|_\beta + Q^\alpha + \Big[\sigma^{\alpha 3} z\Big]_{-h/2}^{+h/2} + \int_{-h/2}^{+h/2} X^\alpha z\,dz = 0.$$

The third term represents the moment of some surface shear stresses $\sigma^{\alpha 3}$ at lever arms $\pm h/2$, and the fourth term is the moment of volume forces acting on the plate material. We may combine them into an external moment $m^\alpha$, the moment per unit area of the middle surface. The equilibrium condition thus assumes its final form

$$M^{\alpha\beta}|_\beta = Q^\alpha + m^\alpha. \tag{7.68}$$

It represents the moment equilibrium of a plate element.

We return to (6.6) and now write it for $i = 3$:

$$\sigma^{3\beta}|_\beta + \sigma^{33}|_3 + X^3 = 0.$$

We multiply by $dz$ and integrate:

$$\int_{-h/2}^{+h/2} \sigma^{3\beta}|_\beta \, dz + \int_{-h/2}^{+h/2} \sigma^{33}|_3 \, dz + \int_{-h/2}^{+h/2} X^3 \, dz = 0.$$

In the second term we may again write $\sigma^{33}|_3 = \sigma^{33}{}_{,3}$ and then integrate to obtain

$$(\sigma^{33})_{z=+h/2} - (\sigma^{33})_{z=-h/2},$$

which is the resultant of normal surface tractions (positive when upward) and may be combined with the volume forces of the third integral into the load $p_3$ (positive when downward) of the plate. The equilibrium condition of vertical forces then reads

$$Q^\alpha|_\alpha + p^3 = 0. \tag{7.69}$$

Equations (7.68) and (7.69) are the equilibrium conditions of the plate element. Together with the elastic law in the form (7.63) or (7.65) they are $2 + 1 + 3 = 6$ equations for three moments $M^{\alpha\beta}$, two shear forces $Q^\alpha$ and the deflection $w$ and, hence, a complete set of equations.

We may now proceed along well-established lines of plate theory to condense these equations into a single differential equation for the deflection $w$. First we differentiate (7.68) with respect to $x^\alpha$ and introduce $Q^\alpha|_\alpha$ from (7.69):

$$M^{\alpha\beta}|_{\alpha\beta} = m^\alpha|_\alpha - p^3. \tag{7.70}$$

Then we use the elastic law to express $M^{\alpha\beta}$ in terms of $w$. For the anisotropic plate, (7.63) leads to

$$\bar{E}^{\alpha\beta\gamma\delta} \frac{h^3}{12} w|_{\alpha\beta\gamma\delta} = p^3 - m^\alpha|_\alpha \tag{7.71}$$

and for the isotropic plate we find from (7.65) the corresponding form of the differential equation:

$$K[(1 - \nu) w|_{\alpha\beta}^{\alpha\beta} + \nu w|_{\gamma\beta}^{\gamma\beta}] = p^3 - m^\alpha|_\alpha.$$

Changing the dummy $\gamma$ into $\alpha$ simplifies this to read

$$K w|_{\alpha\beta}^{\alpha\beta} = p^3 - m^\alpha|_\alpha. \tag{7.72}$$

As explained on page 112,

$$w|_{\alpha\beta}^{\alpha\beta} = \nabla^2\nabla^2 w$$

is the bi-Laplacian of $w$, and (7.72) is the tensor form of the well-known plate equation.

*Problem*

7.1. A plane curvilinear coordinate system $x^\alpha$ is defined by its relation to cartesian coordinates $x, y$:

$$x(\text{Cosh } x^1 + \cos x^2) = c \text{ Sinh } x^1,$$
$$y(\text{Cosh } x^1 + \cos x^2) = c \sin x^2.$$

Sketch a few of the coordinate lines. Transform the plate equation (7.72) into these coordinates. Assume $m^\alpha \equiv 0$.

## References

The plane problems and the torsion problem are standard fare in solution-oriented books on elasticity. For the plane problems see for example [13, pp. 184 and 324], [19, p. 204], [29, p. 249], and [33, p. 15]; for the torsion of prismatic bars [19, p. 310], [29, p. 109], and [33, p. 291]. The book by Timoshenko–Goodier [33] is the most strongly solution-oriented and contains a wide variety of solved problems. The other three books lean more toward problem formulation and general solution methods.

A book by Milne-Thomson [23] is devoted to a peculiar group of problems that can be formulated in two coordinates $x^\alpha$. Our generalized plane strain problem belongs in this group.

For the theory of plate bending see the books by Girkmann [12] and Timoshenko–Woinowsky-Krieger [34]. Girkmann treats also the plane stress problem. Both books are strongly solution-oriented.

CHAPTER 8

# Geometry of Curved Surfaces

ONE OF THE most spectacular applications of tensor calculus in the field of continuum mechanics is the general theory of shells. To study it, we shall need some knowledge of the theory of curved surfaces and we shall now embark upon this mathematical subject, keeping in mind the application we later want to make of it.

## 8.1. General Considerations

On a curved surface it is, in general, not possible to trace a cartesian network. This fact has far-reaching consequences and it is worth while to begin with some general thoughts about the geometry of curved surfaces.

We are living and thinking in a three-dimensional, Euclidean space, in which the curved surface is embedded. We can obtain all necessary geometric information from a three-dimensional analysis, and this will be the way to get into the problem. In the end, however, we want equations valid on the two-dimensional surface. This is a two-dimensional subspace of the general three-dimensional Euclidean space, but it is curved and in it some of the basic concepts of Euclidean geometry do not apply.

To illustrate the situation, let us look at Figure 8.1, which shows one octant of a sphere. On this spherical surface we want to define vectors. Since vectors are usually considered to be elements of straight lines, there is no such thing as a vector on a curved surface, but we may reconcile the contradiction if we restrict ourselves to vectors of infinitesimal length (line elements) or to vectors representing physical quantities like velocity or force. Such vectors have direction and magnitude and they obey the transformation laws, but they do not have any "length" in the geometrical sense, and the

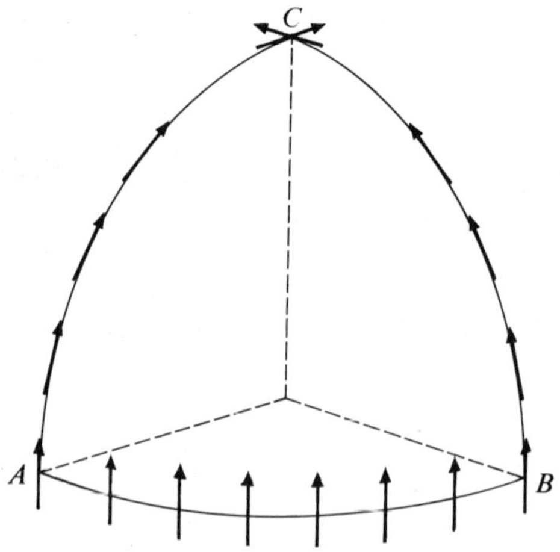


FIGURE 8.1 *Vectors on an octant of a sphere.*

lines which we use to represent them are only geometrical symbols of non-geometrical objects. When we use this modified (or generalized) concept of a vector, it makes sense to speak of vectors on (or in) a curved surface and to ask whether two vectors are equal, that is, whether they have the same magnitude and direction.

The vectors shown along the great circle $AB$ in Figure 8.1 are of equal magnitude and, in three-dimensional space, parallel to each other, hence equal. Now consider another set of vectors along $AB$, each of them normal to the vector shown and pointing from $A$ toward $B$. They are tangents to the circle $AB$ and, in Euclidean, three-dimensional terminology certainly not parallel to each other. However, on the curved surface, they are the only ones which are at right angles to a set of admittedly parallel vectors and we should not hesitate to call them parallel. If we were to follow a sequence of such arrows on the curved surface of the earth, we would certainly say that we are going "always in the same direction." These vectors are as parallel as they possibly can be on the curved surface, but whether we can accept this new definition of parallelism depends on whether or not we can apply it without somewhere running into a contradiction.

To explore this problem, we go from $A$ to $C$ and, at subsequent points of the great circle $AC$, draw vectors which, in this new sense, would be called parallel. They are shown in Figure 8.1. Now we do the same on the great circle $BC$ and thus obtain at $C$ two vectors at right angles to each other, which are both members of a set of allegedly parallel vectors. This shows that the concept of parallel directions does not exist on a spherical surface and that there is no way of saying that two vectors located at *different* points of this surface are equal. The situation is quite different on a cylinder. Any two vectors which are parallel in a plane may still be considered as parallel when this plane is wrapped around a cylinder.

Another well-known peculiarity of spherical geometry is the fact that in a triangle formed by pieces of great circles (that is, geodesics, the straightest lines available on the curved surface) the sum of the angles is not 180°, but more, 270° in the triangle shown in Figure 8.1. Neither of these facts is particularly relevant for our purpose, but later (p. 141) we shall meet with a very important one, which shakes a well-established mathematical prejudice.

## 8.2. Metric and Curvature

In shell theory we shall consider points located on a certain curved surface, called the middle surface, and in its immediate vicinity. We use a coordinate system consisting of two curvilinear coordinates $x^\alpha$ on that middle surface and the normal distance $x^3$ from it. This is a three-dimensional coordinate system and all that we have learned about such systems can be applied.

We shall be interested in comparing quantities defined for points of the middle surface, that is, as functions of $x^\alpha$, with quantities defined for another surface at a small, constant distance $z$ from it. It will be advantageous to use two different notations for related quantities, one for the middle surface $x^3 = 0$ and another for the generic surface $x^3 = z = \text{const}$, as shown in the following table.

| | Middle surface ($z = 0$) | General surface ($z \neq 0$) |
|---|---|---|
| *Geometry* | | |
| position vector | $\mathbf{s}$ | $\mathbf{r}$ |
| line element | $d\mathbf{s}$ | $d\mathbf{r}$ |
| base vectors | $\mathbf{a}_\alpha, \mathbf{a}^\alpha, \mathbf{a}_3 = \mathbf{a}^3$ | $\mathbf{g}_\alpha, \mathbf{g}^\alpha, \mathbf{g}_3 = \mathbf{g}^3$ |
| metric tensor | $a_{\alpha\beta}, a^{\alpha\beta}$ | $g_{\alpha\beta}, g^{\alpha\beta}$ |
| Christoffel symbols | $\Gamma^\alpha_{\beta\gamma}$ | $\bar{\Gamma}^\alpha_{\beta\gamma}$ |
| permutation tensor | $\epsilon_{\alpha\beta}, \epsilon^{\alpha\beta}$ | $\bar{\epsilon}_{\alpha\beta}, \bar{\epsilon}^{\alpha\beta}$ |
| *Deformation* | | |
| displacement | $\mathbf{u} = u_\alpha \mathbf{a}^\alpha + u_3 \mathbf{a}^3$ | $\mathbf{v} = v_\alpha \mathbf{g}^\alpha + v_3 \mathbf{g}^3$ |
| strain tensor | $\varepsilon_{ij}, \varepsilon_{\alpha\beta}$ | $\eta_{ij}, \eta_{\alpha\beta}$ |

For easier reference, this list contains a few quantities that will not be used immediately. All formulas to be derived for $z \neq 0$ are, in the limit, also valid for the middle surface and may be restated in the symbols applicable to it.

To the in-surface base vectors (1.18) is applicable:

$$\mathbf{a}_\alpha = \mathbf{s}_{,\alpha}, \qquad \mathbf{g}_\alpha = \mathbf{r}_{,\alpha}, \tag{8.1}$$

where $\mathbf{s}$ and $\mathbf{r}$ are position vectors emanating from a fixpoint $O$ outside the surface. Since $x^3$ is a linear distance, the third base vector is a unit vector,

$$\mathbf{g}_3 \cdot \mathbf{g}_3 = 1 \tag{8.2}$$

and the symbols $\mathbf{g}_3$, $\mathbf{g}^3$, $\mathbf{a}_3$, and $\mathbf{a}^3$ may all be used interchangeably. The base vectors $\mathbf{g}_\alpha$, $\mathbf{g}_3$ are a monoclinic reference frame; only one of the angles between them is not (or not necessarily) a right angle. Since $\mathbf{g}_3$ is normal to $\mathbf{g}_\alpha$, we have

$$g_{\alpha 3} = \mathbf{g}_\alpha \cdot \mathbf{g}_3 = 0. \tag{8.3}$$

The metric tensor has the following components:

$$[g_{ij}] = \begin{bmatrix} g_{11} & g_{12} & 0 \\ g_{21} & g_{22} & 0 \\ 0 & 0 & 1 \end{bmatrix} \tag{8.4}$$

and its determinant is

$$g = \begin{vmatrix} g_{11} & g_{12} \\ g_{21} & g_{22} \end{vmatrix}. \tag{8.5}$$

The monoclinic character of the base vectors causes some important simplifications in the Christoffel symbols, but they are not as sweeping as those found on page 106, because of the curvature of the surface. From (8.3) it follows by differentiation that

$$\mathbf{g}_{\alpha,\beta} \cdot \mathbf{g}_3 + \mathbf{g}_{3,\beta} \cdot \mathbf{g}_\alpha = 0,$$

whence, with (5.3a) and the symmetry relation (5.6),

$$\overline{\Gamma}_{\alpha\beta 3} = -\overline{\Gamma}_{3\beta\alpha} = -\overline{\Gamma}_{\beta 3\alpha} = \overline{\Gamma}_{\beta\alpha 3} = -\overline{\Gamma}_{3\alpha\beta} = -\overline{\Gamma}_{\alpha 3\beta}. \tag{8.6}$$

Similarly, differentiation of (8.2) yields

$$\mathbf{g}_{3,\alpha} \cdot \mathbf{g}_3 = 0,$$

hence

$$\overline{\Gamma}_{3\alpha 3} = \overline{\Gamma}_{\alpha 33} = 0 \tag{8.7a}$$

and from $\mathbf{g}_{3,3} = 0$ one derives that

$$\overline{\Gamma}_{33\alpha} = \overline{\Gamma}_{333} = 0. \tag{8.7b}$$

Equations (8.7) state that any $\overline{\Gamma}_{ijk}$ vanishes if *more than one* of its subscripts is a 3.

From (8.4) it also follows that

$$g^{\alpha 3} = 0, \qquad g^{33} = 1,$$

and then (5.4b) yields

$$\overline{\Gamma}^3_{3\alpha} = \overline{\Gamma}_{3\alpha\beta}\, g^{\beta 3} + \overline{\Gamma}_{3\alpha 3}\, g^{33} = \overline{\Gamma}_{3\alpha 3} = 0, \tag{8.8a}$$

and, similarly,

$$\overline{\Gamma}^3_{\alpha 3} = \overline{\Gamma}^\alpha_{33} = \overline{\Gamma}^3_{33} = 0. \tag{8.8b}$$

These results may be introduced into (5.11) and (5.13) and lead to the following formulas for the covariant derivative of a vector:

$$\left.\begin{aligned} v_\alpha|_\beta &= v_{\alpha,\beta} - v_\gamma \overline{\Gamma}^\gamma_{\alpha\beta} - v_3 \overline{\Gamma}^3_{\alpha\beta}, \\ v_\alpha|_3 &= v_{\alpha,3} - v_\gamma \overline{\Gamma}^\gamma_{\alpha 3}, \qquad v_3|_\alpha = v_{3,\alpha} - v_\gamma \overline{\Gamma}^\gamma_{3\alpha}; \end{aligned}\right\} \tag{8.9}$$

$$\left.\begin{aligned} v^\alpha|_\beta &= v^\alpha{}_{,\beta} + v^\gamma \overline{\Gamma}^\alpha_{\beta\gamma} + v^3 \overline{\Gamma}^\alpha_{\beta 3}, \\ v^\alpha|_3 &= v^\alpha{}_{,3} + v^\gamma \overline{\Gamma}^\alpha_{\gamma 3}, \qquad v^3|_\alpha = v^3{}_{,\alpha} + v^\gamma \overline{\Gamma}^3_{\alpha\gamma}. \end{aligned}\right\} \tag{8.10}$$

In further formulas, we shall restrict ourselves to the surface $z = 0$ and use the corresponding notation.

A line element $d\mathbf{s}$ in this surface is what we may call a plane vector; it has no component in the direction of $\mathbf{a}_3$:

$$d\mathbf{s} = \mathbf{a}_\alpha\, dx^\alpha.$$

The square of the line element is

$$d\mathbf{s} \cdot d\mathbf{s} = \mathbf{a}_\alpha\, dx^\alpha \cdot \mathbf{a}_\beta\, dx^\beta = a_{\alpha\beta}\, dx^\alpha\, dx^\beta. \tag{8.11}$$

In differential geometry this expression is known as the *first fundamental form* and its coefficients $a_{11}$, $a_{12} = a_{21}$, $a_{22}$ were, in pretensor days, usually denoted by $A$, $B$, $C$.

The normal vector $\mathbf{a}_3$ is a unit vector. Its direction, but not its length, depends on the coordinates $x^\alpha$. Its derivative is, therefore, a plane vector

$$\mathbf{a}_{3,\alpha} = -b_{\alpha\beta}\, \mathbf{a}^\beta \tag{8.12}$$

whence

$$\mathbf{a}_{3,\alpha} \cdot \mathbf{a}_\beta = -\, b_{\alpha\gamma}\, \mathbf{a}^\gamma \cdot \mathbf{a}_\beta = -\, b_{\alpha\gamma}\, \delta^\gamma_\beta = -\, b_{\alpha\beta}. \tag{8.13}$$

The quantities $b_{\alpha\beta}$ are the covariant components of a plane tensor, called the *curvature tensor* of the surface. Before attempting to justify that name, we derive some useful relations.

By differentiating the orthogonality relation

$$\mathbf{a}_\alpha \cdot \mathbf{a}_3 = 0$$

with respect to $x^\beta$ and by using (8.13), we get

$$\mathbf{a}_{\alpha,\beta} \cdot \mathbf{a}_3 = -\mathbf{a}_\alpha \cdot \mathbf{a}_{3,\beta} = b_{\beta\alpha}. \tag{8.14}$$

We rewrite (5.2) in the new notation and have

$$\mathbf{a}_{\alpha,\beta} = \Gamma_{\alpha\beta\gamma}\,\mathbf{a}^\gamma + \Gamma_{\alpha\beta 3}\,\mathbf{a}^3 \tag{8.15}$$

and, hence,

$$b_{\beta\alpha} = \mathbf{a}_{\alpha,\beta} \cdot \mathbf{a}_3 = \Gamma_{\alpha\beta 3}.$$

Since the Christoffel symbols are symmetric with respect to $\alpha$ and $\beta$, $b_{\alpha\beta}$ is also symmetric with respect to $\alpha$ and $\beta$, and, making use of (8.6), we can write

$$b_{\alpha\beta} = \Gamma_{\alpha\beta 3} = \Gamma^3_{\alpha\beta} = -\Gamma_{3\alpha\beta} = -\Gamma_{3\beta\alpha} = -\Gamma_{\alpha 3\beta}. \tag{8.16}$$

From $b_{\alpha\beta}$ we may derive mixed and contravariant components

$$b^\alpha_\beta = b_{\gamma\beta}\,a^{\gamma\alpha}, \qquad b^{\alpha\beta} = b^\alpha_\gamma\,a^{\gamma\beta}. \tag{8.17a, b}$$

From (8.16) we find then that

$$b^\alpha_\beta = -\Gamma_{3\beta\gamma}\,a^{\alpha\gamma} = -\Gamma^\alpha_{3\beta} = -\Gamma^\alpha_{\beta 3} \tag{8.18}$$

and rewriting (5.3b) with $ijk = 3\beta\alpha$ or $\beta 3\alpha$, we find

$$\mathbf{a}_{3,\beta} \cdot \mathbf{a}^\alpha = \mathbf{a}_{\beta,3} \cdot \mathbf{a}^\alpha = -b^\alpha_\beta. \tag{8.19}$$

Equation (8.12) may be written in the alternate form

$$\mathbf{a}_{3,\alpha} = -b^\beta_\alpha \mathbf{a}_\beta, \tag{8.20}$$

from which follows that

$$d\mathbf{a}_3 = \mathbf{a}_{3,\alpha}\,dx^\alpha = -b_{\alpha\beta}\,\mathbf{a}^\beta\,dx^\alpha \tag{8.21}$$

and

$$d\mathbf{a}_3 \cdot d\mathbf{s} = -b_{\alpha\beta}\,\mathbf{a}^\beta\,dx^\alpha \cdot \mathbf{a}_\gamma\,dx^\gamma = -b_{\alpha\beta}\,\delta^\beta_\gamma\;dx^\alpha\,dx^\gamma = -b_{\alpha\beta}\,dx^\alpha\,dx^\beta. \tag{8.22}$$

In differential geometry the last member of this equation is known as the *second fundamental form* and its coefficients $b_{11}$, $b_{12} = b_{21}$, $b_{22}$ are denoted by $E$, $F$, $G$.

Now let us try to visualize the components $b^\alpha_\beta$ of the curvature tensor and thus justify its name. We assume for a moment that only $b^1_1 \neq 0$ and consider a normal section through the surface along an $x^1$-line as shown in Figure 8.2a. Since $|\mathbf{a}_3| = 1$, the length of the vector

$$d\mathbf{a}_3 = \mathbf{a}_{3,1}\,dx^1 = -b^1_1\mathbf{a}_1\,dx^1$$

equals the angle $d\phi$ by which the normal to the surface rotates when we go

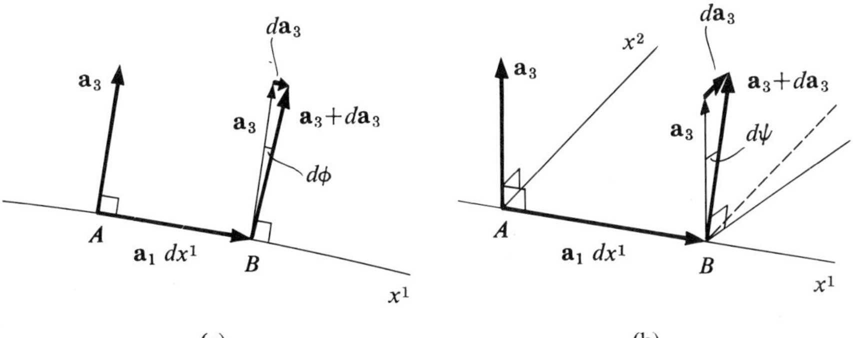


FIGURE 8.2 *Visualization of the curvature components.*

from $A$ to $B$. When we divide this angle by the length $ds$ of the vector $\mathbf{a}_1\, dx^1$, we have the curvature of the surface,

$$\frac{d\phi}{ds} = \frac{|d\mathbf{a}_3|}{|\mathbf{a}_1|\, dx^1} = |b_1^1|.$$

We see that $b_1^1$ is the curvature of the surface in the direction $x^1$.

Now let us assume that $b_1^1 = 0$, but let $b_1^2 \neq 0$. Using Figure 8.2b, we see that

$$d\mathbf{a}_3 = \mathbf{a}_{3,1}\, dx^1 = -b_1^2 \mathbf{a}_2\, dx^1$$

and find for the twist of the surface

$$\frac{d\psi}{ds} = \frac{|d\mathbf{a}_3|}{|\mathbf{a}_1|\, dx^1} = |b_1^2| \frac{|\mathbf{a}_2|}{|\mathbf{a}_1|}.$$

If $|\mathbf{a}_1| = |\mathbf{a}_2|$, then $b_1^2$ and $b_2^1$ are equal to each other and to the twist of the surface with respect to the coordinates $x^\alpha$. If $|\mathbf{a}_1| \neq |\mathbf{a}_2|$, also $b_1^2 \neq b_2^1$, but both are closely related to the twist.

From the curvature tensor two important scalar quantities can be derived—its invariants. The first one is

$$b_\alpha^\alpha = b_1^1 + b_2^2. \tag{8.23}$$

It represents the sum of the two principal curvatures (see p. 179) and is often called the *mean curvature*, although it really is twice that average. The other invariant is the determinant

$$b = \begin{vmatrix} b_1^1 & b_2^1 \\ b_1^2 & b_2^2 \end{vmatrix} = b_1^1 b_2^2 - b_2^1 b_1^2, \tag{8.24}$$

called the *Gaussian curvature* of the surface.

## 8.3. Covariant Derivative

In Section 5.2 we studied the covariant derivative, which describes the rate of change of a vector in a vector field. Now let us consider a vector $\mathbf{v}$ which is defined for all points of the surface as a function of the coordinates $x^\alpha$, but which is not necessarily a plane vector and may have a normal component:

$$\mathbf{v} = v_i \mathbf{a}^i = v_\alpha \mathbf{a}^\alpha + v_3 \mathbf{a}^3, \tag{8.25a}$$

or

$$\mathbf{v} = v^i \mathbf{a}_i = v^\alpha \mathbf{a}_\alpha + v^3 \mathbf{a}_3 . \tag{8.25b}$$

We follow (5.12) and write the derivative of $\mathbf{v}$ with respect to the coordinate $x^\beta$:

$$\mathbf{v}_{,\beta} = v_i|_\beta \mathbf{a}^i = v_\alpha|_\beta \mathbf{a}^\alpha + v_3|_\beta \mathbf{a}^3 \tag{8.26}$$

with

$$v_\alpha|_\beta = v_{\alpha,\beta} - v_\gamma \Gamma^\gamma_{\alpha\beta} - v_3 \Gamma^3_{\alpha\beta},$$

$$v_3|_\beta = v_{3,\beta} - v_\gamma \Gamma^\gamma_{3\beta} - v_3 \Gamma^3_{3\beta}.$$

Making use of (8.8b), (8.16) and (8.18), we can rewrite the last two equations in the following, simpler form

$$v_\alpha|_\beta = v_{\alpha,\beta} - v_\gamma \Gamma^\gamma_{\alpha\beta} - v_3 b_{\alpha\beta}, \tag{8.27}$$

$$v_3|_\beta = v_{3,\beta} + v_\gamma b^\gamma_\beta . \tag{8.28}$$

The first two terms in (8.27) are a two-dimensional counterpart to the covariant derivative defined by (5.13). We need a symbol for it and write

$$v_\alpha\|_\beta = v_{\alpha,\beta} - v_\gamma \Gamma^\gamma_{\alpha\beta}, \tag{8.29}$$

whence

$$v_\alpha|_\beta = v_\alpha\|_\beta - v_3 b_{\alpha\beta} . \tag{8.30}$$

Equation (8.26) may now be rewritten as

$$\begin{aligned}\mathbf{v}_{,\beta} &= (v_\alpha\|_\beta - v_3 b_{\alpha\beta})\mathbf{a}^\alpha + v_3|_\beta \mathbf{a}^3 \\ &= (v_{\alpha,\beta} - v_\gamma \Gamma^\gamma_{\alpha\beta} - v_3 b_{\alpha\beta})\mathbf{a}^\alpha + (v_{3,\beta} + v_\gamma b^\gamma_\beta)\mathbf{a}^3.\end{aligned} \tag{8.31}$$

Similar reasoning, based on (5.10) and (5.11), leads to the definition of the plane covariant derivative of $v^\alpha$:

$$v^\alpha\|_\beta = v^\alpha{}_{,\beta} + v^\gamma \Gamma^\alpha_{\beta\gamma} \tag{8.32}$$

and the relation

$$v^\alpha|_\beta = v^\alpha\|_\beta - v^3 b^\alpha_\beta \tag{8.33}$$

and yields an alternate form of (8.31):

$$\mathbf{v}_{,\beta} = (v^\alpha\|_\beta - v^3 b^\alpha_\beta)\mathbf{a}_\alpha + v^3|_\beta \mathbf{a}_3$$
$$= (v^\alpha{}_{,\beta} + v^\gamma \Gamma^\alpha_{\beta\gamma} - v^3 b^\alpha_\beta)\mathbf{a}_\alpha + (v^3{}_{,\beta} + v^\gamma b_{\gamma\beta})\mathbf{a}_3. \quad (8.34)$$

If $\mathbf{v}$ is not only defined for points on the surface, but also for points adjacent to it, we may speak of a derivative $\mathbf{v}_{,3}$, for which formulas may easily be deduced from (8.9) and (8.10):

$$\mathbf{v}_{,3} = v_\alpha|_3\, \mathbf{a}^\alpha + v_3|_3\, \mathbf{a}^3 = (v_{\alpha,3} + v_\gamma\, b^\gamma_\alpha)\mathbf{a}^\alpha + v_{3,3}\, \mathbf{a}^3$$
$$= v^\alpha|_3\, \mathbf{a}_\alpha + v^3|_3\, \mathbf{a}_3 = (v^\alpha{}_{,3} - v^\gamma b^\alpha_\gamma)\mathbf{a}_\alpha + v^3{}_{,3}\, \mathbf{a}_3. \quad (8.35)$$

There is, of course, no difference between $\mathbf{a}_3$ and $\mathbf{a}^3$ nor between $v_3$ and $v^3$.

In the special case when $\mathbf{v}$ is a plane vector, we have $v_3 = 0$, and there is no difference between the single-dash and double-dash (three-dimensional and two-dimensional) covariant derivatives. Equations (8.31) and (8.34) then simplify to

$$\mathbf{v}_{,\beta} = v_\alpha|_\beta\, \mathbf{a}^\alpha + v_\gamma\, b^\gamma_\beta\, \mathbf{a}^3 = v^\alpha|_\beta\, \mathbf{a}_\alpha + v^\gamma b_{\gamma\beta}\, \mathbf{a}_3 \quad (8.36)$$

and in (8.35) the second term has to be dropped. It is remarkable that $\mathbf{v}_{,\beta}$ has a third component although $\mathbf{v}$ has none.

To establish similar formulas for tensors, we start from (5.23d):

$$A^{ij}|_\gamma = A^{ij}{}_{,\gamma} + A^{\delta j}\Gamma^i_{\gamma\delta} + A^{3j}\Gamma^i_{\gamma 3} + A^{i\,\delta}\Gamma^j_{\gamma\delta} + A^{i3}\Gamma^j_{\gamma 3}. \quad (8.37)$$

If we restrict ourselves to plane tensors with $A^{3\beta} = A^{\alpha 3} = A^{33} = 0$, two terms in (8.37) drop out and what is left is the plane covariant derivative. For $i, j = \alpha, \beta$ we may write

$$A^{\alpha\beta}|_\gamma \equiv A^{\alpha\beta}\|_\gamma = A^{\alpha\beta}{}_{,\gamma} + A^{\delta\beta}\Gamma^\alpha_{\gamma\delta} + A^{\alpha\delta}\Gamma^\beta_{\gamma\delta}. \quad (8.38)$$

However, there are also components $A^{\alpha 3}|_\gamma$ and $A^{3\beta}|_\gamma$, for which we find

$$A^{\alpha 3}|_\gamma = A^{\alpha 3}{}_{,\gamma} + A^{\delta 3}\Gamma^\alpha_{\gamma\delta} + A^{\alpha\delta}\Gamma^3_{\gamma\delta} = A^{\alpha\,\delta} b_{\gamma\delta},$$
$$A^{3\beta}|_\gamma = A^{\delta\beta} b_{\gamma\delta}, \quad (8.39)$$

while $A^{33}|_\gamma = 0$.

From (8.15), which we rewrite in the form

$$\mathbf{a}_{\alpha,\beta} = \Gamma^\delta_{\alpha\beta}\, \mathbf{a}_\delta + b_{\alpha\beta}\, \mathbf{a}_3, \quad (8.40)$$

we may proceed to the second derivative

$$\mathbf{a}_{\alpha,\beta\gamma} = \Gamma^\delta_{\alpha\beta,\gamma}\, \mathbf{a}_\delta + \Gamma^\delta_{\alpha\beta}\, \mathbf{a}_{\delta,\gamma} + b_{\alpha\beta,\gamma}\, \mathbf{a}_3 + b_{\alpha\beta}\, \mathbf{a}_{3,\gamma}.$$

We use (8.20) and (8.40) to eliminate the derivatives of the base vectors:

$$\mathbf{a}_{\alpha,\beta\gamma} = \Gamma^\delta_{\alpha\beta,\gamma}\, \mathbf{a}_\delta + \Gamma^\zeta_{\alpha\beta}(\Gamma^\delta_{\zeta\gamma}\, \mathbf{a}_\delta + b_{\zeta\gamma}\, \mathbf{a}_3) + b_{\alpha\beta,\gamma}\, \mathbf{a}_3 - b_{\alpha\beta}\, b^\delta_\gamma\, \mathbf{a}_\delta$$

and by collecting terms we find

$$\mathbf{a}_{\alpha,\beta\gamma} = (\Gamma^\delta_{\alpha\beta,\gamma} + \Gamma^\zeta_{\alpha\beta}\, \Gamma^\delta_{\zeta\gamma} - b_{\alpha\beta}\, b^\delta_\gamma)\mathbf{a}_\delta + (\Gamma^\zeta_{\alpha\beta}\, b_{\zeta\gamma} + b_{\alpha\beta,\gamma})\mathbf{a}_3. \quad (8.41)$$

This must be equal to the partial derivative $\mathbf{a}_{\alpha,\gamma\beta}$, which we obtain by interchanging $\beta$ and $\gamma$. For the $\mathbf{a}_3$ components the comparison of the second-order derivatives yields the relation

$$b_{\alpha\beta,\gamma} - \Gamma^{\zeta}_{\alpha\gamma} b_{\zeta\beta} = b_{\alpha\gamma,\beta} - \Gamma^{\zeta}_{\alpha\beta} b_{\zeta\gamma}.$$

Since

$$b_{\alpha\beta}\|_{\gamma} = b_{\alpha\beta,\gamma} - b_{\delta\beta}\Gamma^{\delta}_{\alpha\gamma} - b_{\alpha\delta}\Gamma^{\delta}_{\beta\gamma},$$

we see that this may be written as

$$b_{\alpha\beta}\|_{\gamma} = b_{\alpha\gamma}\|_{\beta} \tag{8.42a}$$

or

$$b^{\alpha}_{\beta}\|_{\gamma} = b^{\alpha}_{\gamma}\|_{\beta}. \tag{8.42b}$$

These are two tensor forms of the Gauss–Codazzi equations of differential geometry.

Comparing the coefficient of $\mathbf{a}_\delta$ in (8.41) and its counterpart for $\mathbf{a}_{\alpha,\gamma\beta}$ yields the relation

$$\Gamma^{\delta}_{\alpha\gamma,\beta} - \Gamma^{\delta}_{\alpha\beta,\gamma} + \Gamma^{\zeta}_{\alpha\gamma}\Gamma^{\delta}_{\zeta\beta} - \Gamma^{\zeta}_{\alpha\beta}\Gamma^{\delta}_{\zeta\gamma} = b_{\alpha\gamma} b^{\delta}_{\beta} - b_{\alpha\beta} b^{\delta}_{\gamma}. \tag{8.43}$$

Comparison of the left-hand side of this equation with the coefficient of $v_m$ in (5.28) shows that it is the Riemann–Christoffel tensor in the two-dimensional space of the coordinates $x^\alpha$:

$$R^{\delta}_{.\,\alpha\beta\gamma} = \Gamma^{\delta}_{\alpha\gamma,\beta} - \Gamma^{\delta}_{\alpha\beta,\gamma} + \Gamma^{\zeta}_{\alpha\gamma}\Gamma^{\delta}_{\zeta\beta} - \Gamma^{\zeta}_{\alpha\beta}\Gamma^{\delta}_{\zeta\gamma}, \tag{8.44}$$

and (8.43) states that, on this curved surface, $R^{\delta}_{.\,\alpha\beta\gamma} \neq 0$. This has far-reaching consequences, because, as we have seen on page 73, the order of two subsequent covariant differentiations is not interchangeable unless $R^{\delta}_{.\,\alpha\beta\gamma} = 0$.

If we wish, we may lower the index $\delta$ and, for the Riemann–Christoffel tensor, have the two equivalent statements

$$\left.\begin{aligned} R^{\delta}_{.\,\alpha\beta\gamma} &= b_{\alpha\gamma} b^{\delta}_{\beta} - b_{\alpha\beta} b^{\delta}_{\gamma}, \\ R_{\delta\alpha\beta\gamma} &= b_{\alpha\gamma} b_{\beta\delta} - b_{\alpha\beta} b_{\gamma\delta}. \end{aligned}\right\} \tag{8.45}$$

With the help of (3.29a), the last formula may be further simplified:

$$\begin{aligned} R_{\delta\alpha\beta\gamma} &= b^{\lambda}_{\alpha} b^{\mu}_{\delta}(a_{\gamma\lambda} a_{\beta\mu} - a_{\beta\lambda} a_{\gamma\mu}) = b^{\lambda}_{\alpha} b^{\mu}_{\delta}(\delta^{\nu}_{\lambda}\delta^{\rho}_{\mu} - \delta^{\rho}_{\lambda}\delta^{\nu}_{\mu}) a_{\beta\rho} a_{\gamma\nu} \\ &= b^{\lambda}_{\alpha} b^{\mu}_{\delta} \epsilon^{\rho\nu} \epsilon_{\mu\lambda} a_{\beta\rho} a_{\gamma\nu} = b^{\lambda}_{\alpha} b^{\mu}_{\delta} \epsilon_{\beta\gamma} \epsilon_{\mu\lambda}. \end{aligned}$$

Now we apply the two-dimensional version of the determinant expansion formula (3.17) to the Gaussian curvature $b$:

$$b\epsilon_{\delta\alpha} = b^{\mu}_{\delta} b^{\lambda}_{\alpha} \epsilon_{\mu\lambda},$$

multiply both sides by $\epsilon_{\beta\gamma}$ and find that

$$R_{\delta\alpha\beta\gamma} = b\epsilon_{\delta\alpha}\epsilon_{\beta\gamma}. \tag{8.46}$$

This equation shows that all components of the Riemann–Christoffel tensor $R_{\delta\alpha\beta\gamma}$ vanish if and only if the Gaussian curvature $b = 0$. Then and only then can the sequence of two covariant differentiations be interchanged. The same equation also shows that

$$R_{\delta\alpha\beta\gamma} = -R_{\alpha\delta\beta\gamma} = -R_{\delta\alpha\gamma\beta} = R_{\alpha\delta\gamma\beta}\,. \tag{8.47}$$

If $R_{\delta\alpha\beta\gamma} \neq 0$, then also $v_\alpha\|_{\beta\gamma} \neq v_\alpha\|_{\gamma\beta}$ and we may ask how much they differ. We derive from (5.27) a definition of the second covariant derivative simply by restricting the range of all indices to two dimensions:

$$v_\alpha\|_{\beta\gamma} = (v_{\alpha,\beta} - v_\mu \Gamma^\mu_{\alpha\beta})_{,\gamma} - (v_{\lambda,\beta} - v_\mu \Gamma^\mu_{\lambda\beta})\Gamma^\lambda_{\alpha\gamma} - (v_{\alpha,\lambda} - v_\mu \Gamma^\mu_{\alpha\lambda})\Gamma^\lambda_{\gamma\beta}\,. \tag{8.48}$$

According to (5.28) and (8.46) the difference between this and $v_\alpha\|_{\gamma\beta}$ is

$$v_\alpha\|_{\beta\gamma} - v_\alpha\|_{\gamma\beta} = v^\delta R_{\delta\alpha\beta\gamma} = v^\delta\, b\epsilon_{\delta\alpha}\epsilon_{\beta\gamma}\,. \tag{8.49}$$

Later we shall need a similar relation for the second covariant derivative of a second-order tensor. We start from (5.23b) and apply this equation to a symmetric, two-dimensional tensor $A^{\cdot\delta}_{\alpha} = A^\delta_\alpha$:

$$A^\delta_\alpha\|_\beta = A^\delta_{\alpha,\beta} + A^\zeta_\alpha \Gamma^\delta_{\zeta\beta} - A^\delta_\zeta \Gamma^\zeta_{\alpha\beta}\,.$$

Differentiating now with respect to $x^\gamma$, we find

$$\begin{aligned} A^\delta_\alpha\|_{\beta\gamma} = {} & (A^\delta_{\alpha,\beta} + A^\zeta_\alpha \Gamma^\delta_{\zeta\beta} - A^\delta_\zeta \Gamma^\zeta_{\alpha\beta})_{,\gamma} + (A^\eta_{\alpha,\beta} + A^\zeta_\alpha \Gamma^\eta_{\zeta\beta} - A^\eta_\zeta \Gamma^\zeta_{\alpha\beta})\Gamma^\delta_{\eta\gamma} \\ & - (A^\delta_{\eta,\beta} + A^\zeta_\eta \Gamma^\delta_{\zeta\beta} - A^\delta_\zeta \Gamma^\zeta_{\eta\beta})\Gamma^\eta_{\alpha\gamma} - (A^\delta_{\alpha,\eta} + A^\zeta_\alpha \Gamma^\delta_{\zeta\eta} - A^\delta_\zeta \Gamma^\zeta_{\alpha\eta})\Gamma^\eta_{\beta\gamma}\,. \end{aligned}$$

When we now form the difference $A^\delta_\alpha\|_{\beta\gamma} - A^\delta_\alpha\|_{\gamma\beta}$, very many terms cancel and we are left with

$$\begin{aligned} A^\delta_\alpha\|_{\beta\gamma} - A^\delta_\alpha\|_{\gamma\beta} = {} & A^\zeta_\alpha(\Gamma^\delta_{\zeta\beta,\gamma} - \Gamma^\delta_{\zeta\gamma,\beta} + \Gamma^\delta_{\eta\gamma}\Gamma^\eta_{\zeta\beta} - \Gamma^\delta_{\eta\beta}\Gamma^\eta_{\zeta\gamma}) \\ & + A^\delta_\zeta(\Gamma^\zeta_{\alpha\gamma,\beta} - \Gamma^\zeta_{\alpha\beta,\gamma} + \Gamma^\zeta_{\eta\beta}\Gamma^\eta_{\alpha\gamma} - \Gamma^\zeta_{\eta\gamma}\Gamma^\eta_{\alpha\beta}). \end{aligned}$$

With the help of (8.44) this can be brought into the form

$$A^\delta_\alpha\|_{\beta\gamma} - A^\delta_\alpha\|_{\gamma\beta} = A^\zeta_\alpha R^{\delta}_{\cdot\zeta\gamma\beta} + A^\delta_\zeta R^{\zeta}_{\cdot\alpha\beta\gamma}\,. \tag{8.50}$$

## *Problems*

8.1. On a circular cylinder of radius $a$ a coordinate system is defined as follows (see Figure 8.3):

$$x^1 = z - a\theta \tan\omega = z - c\theta, \qquad x^2 = \theta,$$

where $\omega$ and $c = a\tan\omega$ are constants. Find the metric tensor $a_{\alpha\beta}$ and the curvature tensors $b_{\alpha\beta}$ and $b^\alpha_\beta$.

8.2. In a cartesian coordinate system $x$, $y$, $z$ a hyperbolic paraboloid has the equation $z = xy/c$. The cartesian coordinates $x = x^1$ and $y = x^2$ are used as coordinates $x^\alpha$ of a point on the surface. Calculate, in terms of these $x^\alpha$, the following quantities: $a_{\alpha\beta}$, $a^{\alpha\beta}$, $b_{\alpha\beta}$, $b^\alpha_\beta$, $b^{\alpha\beta}$, $\Gamma_{\alpha\beta\gamma}$, $\Gamma^\gamma_{\alpha\beta}$, $R^{\delta}_{\cdot\alpha\beta\gamma}$.

8.3. On a right circular cone (Figure 8.4a) coordinates $x^\alpha$ are defined as shown. Starting from the expression for the line element, calculate $a_{\alpha\beta}$, $b_{\alpha\beta}$, $\Gamma_{\alpha\beta\gamma}$, $R^{\delta}_{\cdot\alpha\beta\gamma}$.

Do the same for a skew cone (Figure 8.4b).

8.4. Write the Navier–Stokes equation (6.37) as three component equations in the spherical coordinate system of Figure 1.5. Then assume that the radial component of the velocity is zero. This leads to the differential equation of fluid motion on the surface of a sphere.

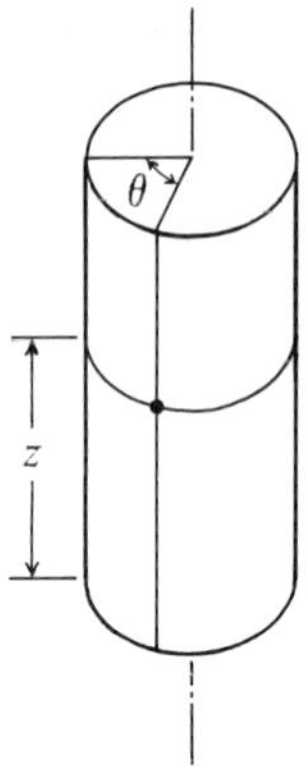


FIGURE 8.3

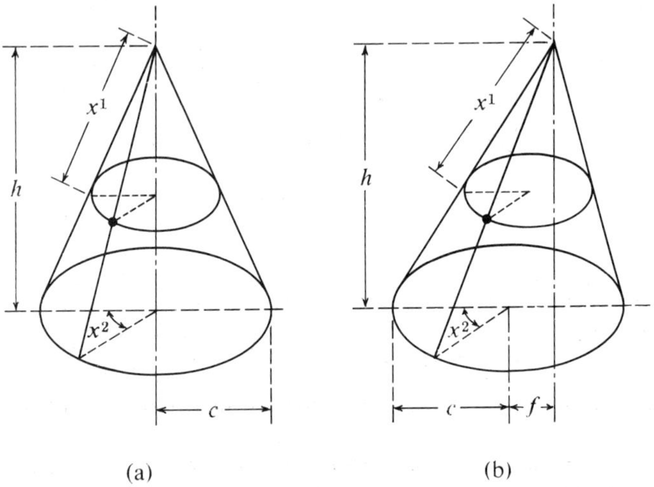


FIGURE 8.4

CHAPTER 9

# Theory of Shells

IN THE PRECEDING CHAPTER we studied the geometry of curved surfaces with the intention of applying the results to the theory of shells. By a shell we understand a piece of solid matter contained in the narrow space between two curved surfaces which are parallel or almost parallel to each other. Their distance is the shell thickness $h$, which is supposed to be small compared with other dimensions of the shell, in particular, with its radii of curvature. The surface which halves the shell thickness everywhere is called the middle surface and serves in the stress analysis the same purpose as the middle plane of a plate or the axis of a beam. The deformation of the shell is described in terms of the deformation of its middle surface, and the stress system is described by stress resultants like those in plates and slabs, referred to a unit length of section through the middle surface.

## 9.1. Shell Geometry

We consider a shell of uniform thickness $h$ and count the coordinate $x^3 = z$ from its middle surface. The outer faces then have the equations $z = \pm h/2$. Now consider a point $B(x^\alpha, x^3)$. Our goal is to express all the quantities associated with this point, like $\mathbf{g}_\alpha$, $g_{\alpha\beta}$, $\overline{\Gamma}_{\alpha\beta\gamma}$, etc., in terms of $z$ and of the corresponding quantities $\mathbf{a}_\alpha$, $a_{\alpha\beta}$, $\Gamma_{\alpha\beta\gamma}$, etc., associated with the midsurface point $A(x^\alpha, 0)$ (Figure 9.1). Both points have the same coordinates $x^\alpha$. Another pair of corresponding points, $D$ and $C$, have the same coordinates $x^\alpha + dx^\alpha$. The plane vectors $AC$ and $BD$ have the same components $dx^\alpha$, but in different sets of base vectors, $\mathbf{a}_\alpha$ and $\mathbf{g}_\alpha$, respectively.

We read from the figure that

$$\mathbf{r} = \mathbf{s} + z\mathbf{a}_3$$

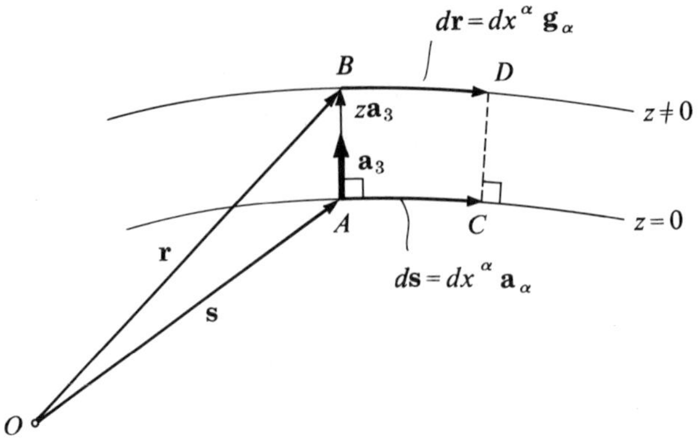


FIGURE 9.1 *Section through a shell.*

and find by differentiation that

$$\mathbf{r}_{,\alpha} = \mathbf{s}_{,\alpha} + z\mathbf{a}_{3,\alpha}$$

and with (1.18) and (8.20):

$$\mathbf{g}_\alpha = \mathbf{a}_\alpha - zb^\gamma_\alpha \mathbf{a}_\gamma = \mu^\gamma_\alpha \mathbf{a}_\gamma \tag{9.1}$$

with

$$\mu^\alpha_\beta = \delta^\alpha_\beta - zb^\alpha_\beta. \tag{9.2}$$

This defines a tensor $\mu^\alpha_\beta$, which relates the base vectors at the points $B$ and $A$. To obtain similar formulas for the contravariant quantities, we introduce a tensor $\lambda^\beta_\alpha$ by tentatively setting

$$\mathbf{g}^\beta = \lambda^\beta_\alpha \mathbf{a}^\alpha. \tag{9.3}$$

Then,

$$\mathbf{g}^\beta \cdot \mathbf{g}_\delta = \lambda^\beta_\alpha \mathbf{a}^\alpha \cdot \mu^\gamma_\delta \mathbf{a}_\gamma = \lambda^\beta_\alpha \mu^\gamma_\delta \, \delta^\alpha_\gamma = \lambda^\beta_\alpha \mu^\alpha_\delta.$$

In connection with (1.20) this shows that

$$\lambda^\beta_\alpha \mu^\alpha_\delta = \delta^\beta_\delta, \tag{9.4}$$

and from this equation the $\lambda^\beta_\alpha$ can be calculated. They are fractions of two polynomials in $z$. For our purposes we need $\lambda^\beta_\alpha$ as a power expansion in $z$, correct to the quadratic term:†

$$\lambda^\beta_\alpha = A^\beta_\alpha + B^\beta_\alpha z + C^\beta_\alpha z^2 + \cdots.$$

† Note that $z$, not being a tensor symbol, will be used with exponents indicating powers.

When this and $\mu_\beta^\alpha$ from (9.2) are introduced into (9.4), we can compare coefficients of like powers of $z$ to obtain $A_\alpha^\beta, \ldots$. This yields the following result:

$$\lambda_\alpha^\beta = \delta_\alpha^\beta + b_\alpha^\beta z + b_\gamma^\beta b_\alpha^\gamma z^2 + \cdots. \tag{9.5}$$

From the definition (1.24) of the metric tensor combined with (9.1) and (9.3) it follows that

$$g_{\alpha\beta} = \mathbf{g}_\alpha \cdot \mathbf{g}_\beta = \mu_\alpha^\gamma \mathbf{a}_\gamma \cdot \mu_\beta^\delta \mathbf{a}_\delta = \mu_\alpha^\gamma \mu_\beta^\delta a_{\gamma\delta}, \tag{9.6a}$$

$$g^{\alpha\beta} = \mathbf{g}^\alpha \cdot \mathbf{g}^\beta = \lambda_\gamma^\alpha \mathbf{a}^\gamma \cdot \lambda_\delta^\beta \mathbf{a}^\delta = \lambda_\gamma^\alpha \lambda_\delta^\beta a^{\gamma\delta}. \tag{9.6b}$$

For later use we also list the following products:

$$\mathbf{g}_\alpha \cdot \mathbf{a}^\beta = \mu_\alpha^\gamma \mathbf{a}_\gamma \cdot \mathbf{a}^\beta = \mu_\alpha^\beta, \tag{9.7a}$$

$$\mathbf{g}^\alpha \cdot \mathbf{a}_\beta = \lambda_\gamma^\alpha \mathbf{a}^\gamma \cdot \mathbf{a}_\beta = \lambda_\beta^\alpha. \tag{9.7b}$$

The tensors $\mu_\alpha^\beta$ and $\lambda_\beta^\alpha$ are of great importance for the development of the shell equations. Therefore, we compile the formulas which will later be needed.

Differentiating (9.4), we find that

$$\lambda_\alpha^\beta\|_\gamma \mu_\delta^\alpha + \lambda_\alpha^\beta \mu_\delta^\alpha\|_\gamma = 0, \tag{9.8}$$

where, of course,

$$\mu_\delta^\alpha\|_\gamma = \mu_{\delta,\gamma}^\alpha + \mu_\delta^\beta \Gamma_{\beta\gamma}^\alpha - \mu_\beta^\alpha \Gamma_{\delta\gamma}^\beta. \tag{9.9}$$

Codazzi's equation (8.42) together with the definition (9.2) yields

$$\mu_\beta^\alpha\|_\gamma = \mu_\gamma^\alpha\|_\beta. \tag{9.10}$$

We shall need the determinant

$$\mu = |\mu_\beta^\alpha| = \begin{vmatrix} \mu_1^1 & \mu_2^1 \\ \mu_1^2 & \mu_2^2 \end{vmatrix} = \mu_1^1 \mu_2^2 - \mu_2^1 \mu_1^2. \tag{9.11}$$

Use of (9.2) shows that this is equal to

$$\mu = 1 - z b_\lambda^\lambda + z^2 b \tag{9.12}$$

with $b$, the Gaussian curvature, as defined by (8.24). We may also apply to $\mu$ the two-dimensional version of the determinant expansion formula (3.17):

$$\epsilon_{\alpha\beta}\,\mu = \epsilon_{\gamma\delta}\,\mu_\alpha^\gamma \mu_\beta^\delta, \tag{9.13}$$

whence, after multiplication by $\epsilon^{\alpha\beta}$ and use of (3.29c), we have

$$2\mu = \epsilon^{\alpha\beta}\epsilon_{\gamma\delta}\,\mu_\alpha^\gamma \mu_\beta^\delta. \tag{9.14}$$

We differentiate both sides and note that the permutation tensor is a constant:

$$2\mu\|_\lambda = \epsilon^{\alpha\beta}\epsilon_{\gamma\delta}(\mu_\alpha^\gamma\|_\lambda \mu_\beta^\delta + \mu_\alpha^\gamma \mu_\beta^\delta\|_\lambda) = 2\epsilon^{\alpha\beta}\epsilon_{\gamma\delta}\,\mu_\alpha^\gamma \mu_\beta^\delta\|_\lambda.$$

From (9.4) and (9.13) we see that

$$\lambda_\delta^\beta \epsilon_{\alpha\beta}\,\mu = \epsilon_{\gamma\zeta}\,\mu_\alpha^\gamma \mu_\beta^\zeta \lambda_\delta^\beta = \epsilon_{\gamma\delta}\,\mu_\alpha^\gamma, \tag{9.15}$$

and when we introduce this in the preceding equation, we find that

$$\mu\|_\lambda = \epsilon^{\alpha\beta}\lambda_\delta^\gamma \epsilon_{\alpha\gamma}\,\mu\mu_\beta^\delta\|_\lambda = \lambda_\delta^\gamma\,\delta_\gamma^\beta\,\mu\mu_\beta^\delta\|_\lambda = \mu\lambda_\delta^\gamma\,\mu_\gamma^\delta\|_\lambda\,. \tag{9.16}$$

We also need the derivative of the determinant $\mu$ in the direction normal to the middle surface. Starting from (9.14), we calculate

$$2\mu_{,3} = \epsilon^{\alpha\beta}\epsilon_{\gamma\delta}(\mu_{\alpha,3}^\gamma\,\mu_\beta^\delta + \mu_{\beta,3}^\delta\,\mu_\alpha^\gamma) = 2\epsilon^{\alpha\beta}\epsilon_{\gamma\delta}\,\mu_{\alpha,3}^\gamma\,\mu_\beta^\delta$$

and, using (9.2) and (9.15),

$$\mu_{,3} = -\epsilon^{\alpha\beta}\epsilon_{\gamma\delta}\,b_\alpha^\gamma\,\mu_\beta^\delta = -\epsilon^{\alpha\beta}b_\alpha^\gamma\lambda_\gamma^\delta\epsilon_{\delta\beta}\,\mu = -\delta_\delta^\alpha\,b_\alpha^\gamma\,\lambda_\gamma^\delta\,\mu,$$

from which ultimately

$$\mu_{,3} = -b_\alpha^\gamma\,\lambda_\gamma^\alpha\,\mu. \tag{9.17}$$

We may now derive expressions for the Christoffel symbols $\bar{\Gamma}_{\alpha\beta\gamma}$, $\bar{\Gamma}_{\alpha\beta}^\gamma$ and the permutation tensor $\bar{\epsilon}_{\alpha\beta}$, $\bar{\epsilon}^{\alpha\beta}$ at point $B$ of Figure 9.1 in terms of the corresponding quantities $\Gamma_{\alpha\beta\gamma}$, $\Gamma_{\alpha\beta}^\gamma$, $\epsilon_{\alpha\beta}$, $\epsilon^{\alpha\beta}$ for the point $A$. The former quantities are derived from the base vectors $\mathbf{g}_\alpha$ and the metric tensor $g_{\alpha\beta}$ of point $B$, while the quantities without bars are defined in terms of $\mathbf{a}_\alpha$ and $a_{\alpha\beta}$.

We start from (5.3b), which we rewrite for the two-dimensional range, and use (9.1) and (9.3):

$$\bar{\Gamma}_{\alpha\beta}^\gamma = \mathbf{g}_{\alpha,\beta}\cdot\mathbf{g}^\gamma = (\mu_\alpha^\delta\mathbf{a}_\delta)_{,\beta}\cdot\lambda_\zeta^\gamma\mathbf{a}^\zeta.$$

Now, with an alternate version of (8.15) and with (9.9) we find

$$(\mu_\alpha^\delta\mathbf{a}_\delta)_{,\beta}\cdot\mathbf{a}^\zeta = \mu_{\alpha,\beta}^\delta\,\delta_\delta^\zeta + \mu_\alpha^\delta\,\Gamma_{\delta\beta}^\gamma\,\delta_\gamma^\zeta = \mu_{\alpha,\beta}^\zeta + \mu_\alpha^\delta\,\Gamma_{\delta\beta}^\zeta = \mu_\alpha^\zeta\|_\beta + \mu_\delta^\zeta\,\Gamma_{\alpha\beta}^\delta$$

and hence

$$\bar{\Gamma}_{\alpha\beta}^\gamma = \lambda_\zeta^\gamma\,\mu_\alpha^\zeta\|_\beta + \delta_\delta^\gamma\,\Gamma_{\alpha\beta}^\delta = \Gamma_{\alpha\beta}^\gamma + \lambda_\zeta^\gamma\,\mu_\alpha^\zeta\|_\beta\,. \tag{9.18}$$

Similarly we find with (8.15)

$$\bar{\Gamma}_{\alpha\beta}^3 = \mathbf{g}_{\alpha,\beta}\cdot\mathbf{g}^3 = (\mu_\alpha^\delta\,\mathbf{a}_\delta)_{,\beta}\cdot\mathbf{a}^3 = \mu_\alpha^\delta\,b_{\delta\beta} \tag{9.19}$$

and with (8.12) and $\mathbf{g}_3 = \mathbf{a}_3$:

$$\bar{\Gamma}_{3\beta}^\alpha = \mathbf{g}_{3,\beta}\cdot\mathbf{g}^\alpha = -\,b_{\beta\gamma}\,\mathbf{a}^\gamma\cdot\lambda_\delta^\alpha\mathbf{a}^\delta = -\lambda_\delta^\alpha\,b_{\beta\gamma}\,a^{\gamma\delta} = -\lambda_\delta^\alpha\,b_\beta^\delta\,. \tag{9.20}$$

For the permutation tensor, we start from

$$\mathbf{g}_\beta\times\mathbf{g}_3 = \bar{\epsilon}_{\beta 3\delta}\,\mathbf{g}^\delta = \bar{\epsilon}_{\delta\beta}\,\mathbf{g}^\delta$$

and use (9.1) and (9.3):

$$\mu_\beta^\delta\,\mathbf{a}_\delta\times\mathbf{a}_3 = \bar{\epsilon}_{\delta\beta}\,\lambda_\gamma^\delta\,\mathbf{a}^\gamma.$$

The left-hand side equals

$$\mu_\beta^\delta \epsilon_{\gamma\delta} \mathbf{a}^\gamma.$$

Now, when we equate the $\gamma$-components on both sides and multiply them by $\mu_\alpha^\gamma$, we have

$$\mu_\alpha^\gamma \mu_\beta^\delta \epsilon_{\gamma\delta} = \bar{\epsilon}_{\delta\beta} \delta_\alpha^\delta,$$

whence

$$\bar{\epsilon}_{\alpha\beta} = \mu_\alpha^\gamma \mu_\beta^\delta \epsilon_{\gamma\delta}. \tag{9.21}$$

Comparison of the right-hand side of this equation with that of (9.13) shows that

$$\bar{\epsilon}_{\alpha\beta} = \epsilon_{\alpha\beta} \mu. \tag{9.22}$$

Both expressions will later be useful, each one in its place.

## 9.2. Kinematics of Deformation

When a shell is deformed, the position vector $\mathbf{r}$ of a generic point $B$ (Figure 9.2) with coordinates $x^\alpha$, $x^3$ changes to

$$\hat{\mathbf{r}} = \mathbf{r} + \mathbf{v}, \tag{9.23}$$

while that of the corresponding point $A$ on the middle surface changes from $\mathbf{s}$ to

$$\hat{\mathbf{s}} = \mathbf{s} + \mathbf{u}. \tag{9.24}$$

Since we associate with the same material point before and after deformation the same coordinates $x^\alpha$, $x^3$, the coordinate net is deformed with the body and we have, in the deformed state, new base vectors

$$\hat{\mathbf{g}}_\alpha = \hat{\mathbf{r}}_{,\alpha} \qquad \text{and} \qquad \hat{\mathbf{a}}_\alpha = \hat{\mathbf{s}}_{,\alpha}. \tag{9.25a,b}$$

We set the goal of expressing these quantities as well as the displacement $\mathbf{v}$ and the components of the strain tensor $\eta_{ij}$ at $B$ in terms of the displacement $\mathbf{u}$ of point $A$ and its derivatives. In this work we use the fundamental assumption of the *conservation of normals*; that is, we assume that all particles lying on a normal to the undeformed middle surface will, after deformation, be found on a normal to the deformed middle surface. This amounts to assuming that the components $\eta_{\alpha 3} = \eta_{3\alpha}$ of the strain tensor $\eta_{ij}$ are zero.

The relation between strain and displacement is given by (6.1), which we rewrite in the notation for a point off the middle surface (see p. 133):

$$\eta_{ij} = \tfrac{1}{2}(v_i|_j + v_j|_i). \tag{9.26}$$

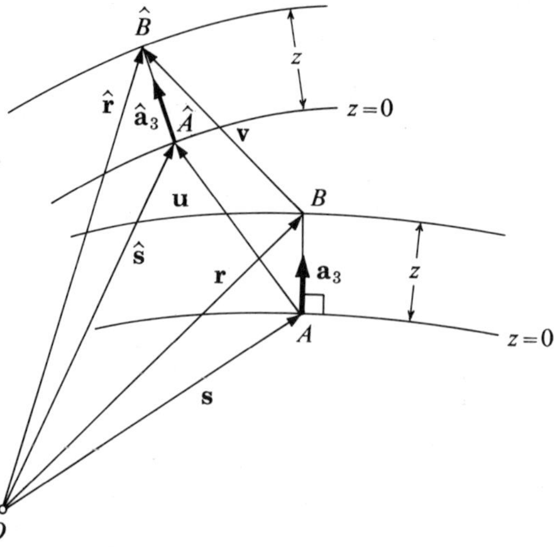


FIGURE 9.2 *Section through a shell before and after deformation.*

We let $i = \alpha, j = 3$ and apply (8.9):

$$\eta_{\alpha 3} = \tfrac{1}{2}(v_{\alpha,3} - v_\beta \overline{\Gamma}^\beta_{\alpha 3} + v_{3,\alpha} - v_\beta \overline{\Gamma}^\beta_{3\alpha}).$$

From the conservation of normals we conclude that

$$v_{\alpha,3} + v_{3,\alpha} = 2v_\beta \overline{\Gamma}^\beta_{3\alpha}. \tag{9.27}$$

Since the shell is thin and since there is no stress $\sigma^{33}$ in the direction of the base vector $\mathbf{g}_3$, the length of the normals does not change much during deformation, so that we may set

$$v_3 = u_3 = w. \tag{9.28}$$

This statement is not part of the assumption of the conservation of normals and must be used with some care. It is good for the kinematic purposes for which we shall use it and amounts there essentially to a linearization in the displacements. However, the statement $\varepsilon_{33} = 0$, which this assumption implies, must not be inserted into Hooke's law, which then would yield a substantial $\sigma^{33}$. The shell is definitely not in a state of plane strain, but rather of plane stress including the stress system known from the bending of plates.

When (9.27) is applied to the middle surface, (8.18) may be used to introduce the curvature tensor

$$u_{\alpha,3} + w_{,\alpha} = -2u_\beta b^\beta_\alpha. \tag{9.29}$$

Our next aim is to get $\hat{\mathbf{a}}_3$, the normal vector of a deformed shell element. We differentiate (9.23) with respect to $x^3$ and then let $z = 0$, writing $\mathbf{s}_{,i}$ for $\lim \mathbf{r}_{,i}$. This yields

$$\hat{\mathbf{s}}_{,3} = \mathbf{s}_{,3} + \mathbf{u}_{,3}$$

and with (1.18) and (8.35)

$$\hat{\mathbf{a}}_3 = \mathbf{a}_3 + (u_{\alpha,3} + u_\gamma b_\alpha^\gamma)\mathbf{a}^\alpha + u_{3,3}\,\mathbf{a}^3.$$

Since $u_3 = w$ does not depend on $x^3$, we have

$$\hat{\mathbf{a}}_3 - \mathbf{a}_3 = (u_{\alpha,3} + u_\gamma b_\alpha^\gamma)\mathbf{a}^\alpha, \tag{9.30}$$

and because of (9.29) this may be written in the alternate form

$$\hat{\mathbf{a}}_3 - \mathbf{a}_3 = -(w_{,\alpha} + u_\gamma b_\alpha^\gamma)\mathbf{a}^\alpha. \tag{9.31}$$

Since both $\hat{\mathbf{a}}_3$ and $\mathbf{a}_3$ are unit vectors, their difference is the angle of rotation of the normal. It is a plane vector and (9.31) confirms that the partial derivatives $w_{,\alpha}$ of the scalar $w$ are vector components. This permits us to use the notation

$$\omega_\alpha = w_{,\alpha} = w|_\alpha = w\|_\alpha. \tag{9.32}$$

Let us note in passing that, from (8.29),

$$w\|_{\alpha\beta} = \omega_\alpha\|_\beta = w_{,\alpha\beta} - w_{,\gamma}\Gamma_{\alpha\beta}^\gamma = w\|_{\beta\alpha}, \tag{9.33}$$

a relation that will be of use later.

Once more, we use Figure 9.2, which shows two points $A$ and $B$ before deformation and their positions $\hat{A}$ and $\hat{B}$ after deformation. From this figure we read that

$$\hat{\mathbf{r}} = \mathbf{r} + \mathbf{v} = \mathbf{s} + z\mathbf{a}_3 + \mathbf{v}$$

and also

$$\hat{\mathbf{r}} = \hat{\mathbf{s}} + z\hat{\mathbf{a}}_3 = \mathbf{s} + \mathbf{u} + z\hat{\mathbf{a}}_3.$$

We equate the right-hand sides and use (9.31):

$$\mathbf{v} = \mathbf{u} + z(\hat{\mathbf{a}}_3 - \mathbf{a}_3) = u_\delta \mathbf{a}^\delta + w\mathbf{a}^3 - z(w_{,\delta} + u_\gamma b_\delta^\gamma)\mathbf{a}^\delta. \tag{9.34}$$

On the other hand, we have

$$\mathbf{v} = v_\gamma \mathbf{g}^\gamma + w\mathbf{g}^3$$

and hence

$$\mathbf{v} \cdot \mathbf{g}_\alpha = v_\gamma \mathbf{g}^\gamma \cdot \mathbf{g}_\alpha = v_\gamma \delta_\alpha^\gamma = v_\alpha.$$

We apply this to (9.34) and find the component

$$v_\alpha = [u_\delta - z(w_{,\delta} + u_\gamma b_\delta^\gamma)]\mathbf{a}^\delta \cdot \mathbf{g}_\alpha.$$

Because of (9.2) and (9.7a), this is equivalent to

$$v_\alpha = (u_\gamma \mu^\gamma_\delta - z w_{,\delta}) \mu^\delta_\alpha . \tag{9.35}$$

To calculate the strain $\eta_{\alpha\beta}$ from (9.26), we need the covariant derivative $v_\alpha|_\beta$. This derivative is to be taken using the metric at point $B$, i.e. with the Christoffel symbols derived from $\mathbf{g}_\alpha$, $\mathbf{g}_3$:

$$v_\alpha|_\beta = v_{\alpha,\beta} - v_\rho \overline{\Gamma}^\rho_{\alpha\beta} - v_3 \overline{\Gamma}^3_{\alpha\beta} .$$

We apply (9.18) and (9.19):

$$v_\alpha|_\beta = v_{\alpha,\beta} - v_\rho(\Gamma^\rho_{\alpha\beta} + \lambda^\rho_\delta \mu^\delta_\alpha\|_\beta) - w\mu^\gamma_\alpha b_{\gamma\beta} = v_\alpha\|_\beta - v_\rho \lambda^\rho_\delta \mu^\delta_\alpha\|_\beta - w\mu^\gamma_\alpha b_{\gamma\beta} .$$

Now it is time to introduce (9.35) and to express $v_\alpha$, $v_\rho$ in terms of the displacement of the middle surface:

$$\begin{aligned} v_\alpha|_\beta &= [(u_\gamma \mu^\gamma_\delta - z w\|_\delta)\mu^\delta_\alpha]\|_\beta - (u_\gamma \mu^\gamma_\sigma - z w\|_\sigma)\mu^\sigma_\rho \lambda^\rho_\delta \mu^\delta_\alpha\|_\beta - w\mu^\gamma_\alpha b_{\gamma\beta} \\ &= u_\gamma\|_\beta \mu^\gamma_\delta \mu^\delta_\alpha + u_\gamma \mu^\gamma_\delta\|_\beta \mu^\delta_\alpha + u_\gamma \mu^\gamma_\delta \mu^\delta_\alpha\|_\beta - z(w\|_{\delta\beta} \mu^\delta_\alpha + w\|_\delta \mu^\delta_\alpha\|_\beta) \\ &\quad - u_\gamma \mu^\gamma_\sigma \delta^\sigma_\delta \mu^\delta_\alpha\|_\beta + z w\|_\sigma \delta^\sigma_\delta \mu^\delta_\alpha\|_\beta - w\mu^\gamma_\alpha b_{\gamma\beta} . \end{aligned}$$

Several terms cancel each other and there remains only

$$v_\alpha|_\beta = u_\gamma\|_\beta \mu^\gamma_\delta \mu^\delta_\alpha + u_\gamma \mu^\gamma_\delta\|_\beta \mu^\delta_\alpha - z w\|_{\delta\beta} \mu^\delta_\alpha - w\mu^\gamma_\alpha b_{\gamma\beta} . \tag{9.36}$$

From this, $v_\beta|_\alpha$ is obtained by interchanging $\alpha$ and $\beta$, and then (9.26) yields for the strain

$$\begin{aligned} 2\eta_{\alpha\beta} = v_\alpha|_\beta + v_\beta|_\alpha &= (u_\gamma\|_\alpha \mu^\delta_\beta + u_\gamma\|_\beta \mu^\delta_\alpha)\mu^\gamma_\delta + u_\gamma(\mu^\delta_\alpha \mu^\gamma_\delta\|_\beta + \mu^\delta_\beta \mu^\gamma_\delta\|_\alpha) \\ &\quad - z(w\|_{\alpha\delta} \mu^\delta_\beta + w\|_{\beta\delta} \mu^\delta_\alpha) - w(\mu^\gamma_\alpha b_{\gamma\beta} + \mu^\gamma_\beta b_{\gamma\alpha}). \end{aligned} \tag{9.37}$$

Because of the basic assumption that the normals are conserved, the deformation of a shell element depends exclusively on the deformation of the middle surface and it must be possible to express $\eta_{\alpha\beta}$ in terms of the strains $\varepsilon_{\gamma\delta}$ of the middle surface and of its change of curvature.

The strain of the middle surface is easily obtained from (9.37) by letting $z = 0$. Then the third term drops out and in all the other ones the various $\mu^\eta_\zeta$ become Kronecker deltas with the same indices. All that remains is

$$2\varepsilon_{\alpha\beta} = u_\beta\|_\alpha + u_\alpha\|_\beta - 2wb_{\alpha\beta} . \tag{9.38}$$

Transcribing (2.5) and (2.17) into the present notation, we may then calculate the metric tensor for the deformed middle surface,

$$\hat{a}_{\alpha\beta} = a_{\alpha\beta} + 2\varepsilon_{\alpha\beta} . \tag{9.39}$$

The change of curvature of the middle surface is, of course, the difference between the curvature tensors before and after deformation. Since their components are defined with respect to different reference frames, $\mathbf{a}_\alpha$ and $\hat{\mathbf{a}}_\alpha$, the differences $\hat{b}_{\alpha\beta} - b_{\alpha\beta}$, $\hat{b}^\alpha_\beta - b^\alpha_\beta$, and $\hat{b}^{\alpha\beta} - b^{\alpha\beta}$ are not components of the same tensor and we must choose among them for our definition. We decide in favor of the components of mixed variance, because in such components the metric tensor is independent of the deformation, $\hat{a}^\gamma_\alpha = a^\gamma_\alpha = \delta^\gamma_\alpha$ [see (1.53)]. We define

$$\kappa^\gamma_\alpha = \hat{b}^\gamma_\alpha - b^\gamma_\alpha \tag{9.40}$$

and then calculate

$$\kappa_{\alpha\beta} = \kappa^\gamma_\alpha a_{\gamma\beta} = \hat{b}^\gamma_\alpha a_{\gamma\beta} - b_{\alpha\beta}. \tag{9.41}$$

This tensor is not symmetric because, in the third member,

$$\hat{b}^\gamma_\alpha a_{\gamma\beta} \neq \hat{b}^\gamma_\beta a_{\gamma\alpha}.$$

By introducing $a_{\gamma\beta}$ from (9.39) in (9.41), we find that

$$\kappa_{\alpha\beta} = \hat{b}^\gamma_\alpha(\hat{a}_{\gamma\beta} - 2\varepsilon_{\gamma\beta}) - b_{\alpha\beta} = \hat{b}_{\alpha\beta} - b_{\alpha\beta} - 2\hat{b}^\gamma_\alpha \varepsilon_{\gamma\beta}.$$

If we neglect a term quadratic in the strains, we may write this as

$$\kappa_{\alpha\beta} = \hat{b}_{\alpha\beta} - b_{\alpha\beta} - 2b^\gamma_\alpha \varepsilon_{\gamma\beta}. \tag{9.42}$$

This equation shows that $\kappa_{\alpha\beta}$ derived from our definition of $\kappa^\gamma_\alpha$ is not equal to $\hat{b}_{\alpha\beta} - b_{\alpha\beta}$.

To calculate $\hat{b}_{\alpha\beta}$, we apply (8.13) to the deformed middle surface:

$$\hat{b}_{\alpha\beta} = -\hat{\mathbf{a}}_{3,\alpha} \cdot \hat{\mathbf{a}}_\beta, \tag{9.43}$$

Differentiating (9.31) and using (8.36), we find

$$\hat{\mathbf{a}}_{3,\alpha} = \mathbf{a}_{3,\alpha} - (w\|_\gamma + u_\delta b^\delta_\gamma)\|_\alpha \mathbf{a}^\gamma - (w\|_\gamma + u_\delta b^\delta_\gamma) b^\gamma_\alpha \mathbf{a}^3.$$

To find the deformed base vector $\hat{\mathbf{a}}_\beta$, we start from its definition (9.25b) and use (9.24) and (8.31):

$$\hat{\mathbf{a}}_\beta = \mathbf{s}_{,\beta} + \mathbf{u}_{,\beta} = \mathbf{a}_\beta + (u_\gamma\|_\beta - w b_{\gamma\beta})\mathbf{a}^\gamma + (w_{,\beta} + u_\gamma b^\gamma_\beta)\mathbf{a}^3.$$

When we dot-multiply $\hat{\mathbf{a}}_{3,\alpha}$ and $\hat{\mathbf{a}}_\beta$, we keep only the terms linear in the displacements and use (8.13) and (8.19):

$$\begin{aligned} \hat{b}_{\alpha\beta} &= -\mathbf{a}_{3,\alpha} \cdot \mathbf{a}_\beta - (u_\gamma\|_\beta - w b_{\gamma\beta})\mathbf{a}^\gamma \cdot \mathbf{a}_{3,\alpha} + (w\|_\gamma + u_\delta b^\delta_\gamma)\|_\alpha \mathbf{a}^\gamma \cdot \mathbf{a}_\beta \\ &= b_{\alpha\beta} + (u_\gamma\|_\beta - w b_{\gamma\beta}) b^\gamma_\alpha + w\|_{\beta\alpha} + (u_\delta b^\delta_\beta)\|_\alpha. \end{aligned} \tag{9.44}$$

Equation (9.42) now yields the desired result:

$$\begin{aligned} \kappa_{\alpha\beta} &= (u_\gamma\|_\beta - w b_{\gamma\beta}) b^\gamma_\alpha + w\|_{\alpha\beta} + (u_\delta b^\delta_\beta)\|_\alpha - 2b^\gamma_\alpha \varepsilon_{\gamma\beta} \\ &= u_\gamma b^\gamma_\beta\|_\alpha + u_\gamma\|_\alpha b^\gamma_\beta - u_\beta\|_\gamma b^\gamma_\alpha + w\|_{\alpha\beta} + w b^\gamma_\alpha b_{\gamma\beta}. \end{aligned} \tag{9.45}$$

In the first, fourth, and fifth terms $\alpha$ and $\beta$ can be interchanged because of the

Codazzi equation (8.42b), because of (9.33), and because of (1.22), which applies to any product with one dummy index. In the second and third terms, however, interchanging $\alpha$ and $\beta$ changes the values and, therefore, $\kappa_{\alpha\beta} \neq \kappa_{\beta\alpha}$.

We may now come back to our intention of expressing $\eta_{\alpha\beta}$ in terms of $\varepsilon_{\gamma\delta}$ and $\kappa_{\gamma\delta}$. It seems reasonable to project $\varepsilon_{\gamma\delta}$ into the metric of point $B$ in Figure 9.1 by multiplying it by two $\mu$ factors and, hence, to consider the difference

$$\begin{aligned} 2(\eta_{\alpha\beta} - \varepsilon_{\gamma\delta}\mu_\alpha^\gamma\mu_\beta^\delta) &= u_\gamma\|_\alpha \mu_\delta^\gamma\mu_\beta^\delta + u_\gamma\|_\beta \mu_\delta^\gamma\mu_\alpha^\delta - u_\gamma\|_\delta \mu_\alpha^\gamma\mu_\beta^\delta - u_\delta\|_\gamma \mu_\alpha^\gamma\mu_\beta^\delta \\ &\quad + u_\zeta(\mu_\alpha^\delta\mu_\delta^\zeta\|_\beta + \mu_\beta^\delta\mu_\delta^\zeta\|_\alpha) - w(\mu_\alpha^\gamma b_{\gamma\beta} + \mu_\beta^\gamma b_{\gamma\alpha} - 2\mu_\alpha^\gamma\mu_\beta^\delta b_{\gamma\delta}) \\ &\quad - z(w\|_{\alpha\gamma}\mu_\beta^\gamma + w\|_{\beta\gamma}\mu_\alpha^\gamma). \end{aligned} \tag{9.46}$$

We expect that this equals the product of $z\kappa_{\gamma\delta}$ and some suitable factor and, therefore, try to extract a factor $z$ from every term. In the last term this has already been achieved and all we do is to write it in the slightly changed form

$$-zw\|_{\gamma\delta}(\mu_\alpha^\gamma\,\delta_\beta^\delta + \mu_\beta^\gamma\,\delta_\alpha^\delta).$$

In the other terms it is necessary to make use of (9.2), which defines $\mu_\beta^\alpha$, and of two relations which can easily be derived from it. First, since $\delta_\beta^\alpha$ is a constant, we see that

$$\mu_\beta^\alpha\|_\gamma = -zb_\beta^\alpha\|_\gamma\,; \tag{9.47a}$$

and, second, we find that

$$\mu_\beta^\alpha b_\gamma^\beta = (\delta_\beta^\alpha - zb_\beta^\alpha)b_\gamma^\beta = b_\gamma^\alpha - zb_\beta^\alpha b_\gamma^\beta = b_\beta^\alpha(\delta_\gamma^\beta - zb_\gamma^\beta) = b_\beta^\alpha\mu_\gamma^\beta. \tag{9.47b}$$

We use (9.47a) when dealing with the coefficient of $u_\zeta$ in (9.46):

$$\mu_\alpha^\delta\mu_\delta^\zeta\|_\beta + \mu_\beta^\delta\mu_\delta^\zeta\|_\alpha = -z(\mu_\alpha^\delta b_\delta^\zeta\|_\beta + \mu_\beta^\delta b_\delta^\zeta\|_\alpha) = -zb_\delta^\zeta\|_\gamma(\mu_\alpha^\delta\,\delta_\beta^\gamma + \mu_\beta^\delta\,\delta_\alpha^\gamma).$$

The coefficient of $w$ makes only slightly more work. We use (9.2) and write

$$\begin{aligned} &\mu_\alpha^\gamma b_{\gamma\beta} + \mu_\beta^\gamma b_{\gamma\alpha} - 2\mu_\alpha^\gamma\mu_\beta^\delta b_{\gamma\delta} \\ &\qquad = (\delta_\alpha^\gamma - zb_\alpha^\gamma)b_{\gamma\beta} + (\delta_\beta^\gamma - zb_\beta^\gamma)b_{\gamma\alpha} - 2(\delta_\alpha^\gamma - z\,b_\alpha^\gamma)(\delta_\beta^\delta - zb_\beta^\delta)b_{\gamma\delta}. \end{aligned}$$

In this expression most terms cancel and all that is left is

$$z\delta_\alpha^\gamma b_\beta^\delta\, b_{\gamma\delta} + zb_\alpha^\gamma\,\delta_\beta^\delta\, b_{\gamma\delta} - 2z^2b_\alpha^\gamma b_\beta^\delta\, b_{\gamma\delta} = z(\mu_\alpha^\gamma b_\beta^\delta + \mu_\beta^\delta b_\alpha^\gamma)b_{\gamma\delta}\,,$$

which may be written in the form

$$zb_\gamma^\zeta\, b_{\zeta\delta}(\mu_\alpha^\delta\,\delta_\beta^\gamma + \mu_\beta^\delta\,\delta_\alpha^\gamma).$$

The terms of (9.46) which contain derivatives of $u_\gamma$, may be processed as follows:

$$u_\gamma\|_\alpha \mu^\gamma_\delta \mu^\delta_\beta + u_\gamma\|_\beta \mu^\gamma_\delta \mu^\delta_\alpha - u_\gamma\|_\zeta \mu^\zeta_\beta \mu^\gamma_\delta \delta^\delta_\alpha - u_\gamma\|_\zeta \mu^\zeta_\alpha \mu^\gamma_\delta \delta^\delta_\beta$$
$$= \mu^\gamma_\delta [u_\gamma\|_\alpha(\delta^\delta_\beta - zb^\delta_\beta) + u_\gamma\|_\beta(\delta^\delta_\alpha - zb^\delta_\alpha) - u_\gamma\|_\zeta(\delta^\zeta_\beta - zb^\zeta_\beta)\,\delta^\delta_\alpha - u_\gamma\|_\zeta(\delta^\zeta_\alpha - zb^\zeta_\alpha)\delta^\delta_\beta]$$
$$= -zu_\gamma\|_\zeta \mu^\gamma_\delta[\delta^\zeta_\alpha b^\delta_\beta + \delta^\zeta_\beta b^\delta_\alpha - \delta^\delta_\alpha b^\zeta_\beta - \delta^\delta_\beta b^\zeta_\alpha].$$

We make use of (9.47b) and continue the calculation:

$$= -zu_\gamma\|_\zeta[b^\gamma_\delta(\delta^\zeta_\alpha \mu^\delta_\beta + \delta^\zeta_\beta \mu^\delta_\alpha) - b^\zeta_\delta(\delta^\delta_\beta \mu^\gamma_\alpha + \delta^\delta_\alpha \mu^\gamma_\beta)]$$
$$= -z[u_\zeta\|_\gamma b^\zeta_\delta(\delta^\gamma_\alpha \mu^\delta_\beta + \delta^\gamma_\beta \mu^\delta_\alpha) - u_\delta\|_\zeta\, b^\zeta_\gamma(\delta^\gamma_\beta \mu^\delta_\alpha + \delta^\gamma_\alpha \mu^\delta_\beta)]$$
$$= -z(u_\zeta\|_\gamma b^\zeta_\delta - u_\delta\|_\zeta b^\zeta_\gamma)(\delta^\gamma_\alpha \mu^\delta_\beta + \delta^\gamma_\beta \mu^\delta_\alpha).$$

Now we discover that all terms of (9.46) do not only contain a factor $z$, but still another common factor, and we may rewrite that equation in the following form:

$$2(\eta_{\alpha\beta} - \varepsilon_{\gamma\delta}\mu^\gamma_\alpha \mu^\delta_\beta)$$
$$= z(\delta^\gamma_\alpha \mu^\delta_\beta + \delta^\gamma_\beta \mu^\delta_\alpha)[-u_\zeta\|_\gamma b^\zeta_\delta + u_\delta\|_\zeta b^\zeta_\gamma - u_\zeta b^\zeta_\delta\|_\gamma - wb^\zeta_\gamma b_{\zeta\delta} - w\|_{\gamma\delta}].$$

Comparison of the bracketed expression with the third member of (9.45) shows that it equals $-\kappa_{\gamma\delta}$. Therefore, we can write

$$\eta_{\alpha\beta} = \varepsilon_{\gamma\delta}\mu^\gamma_\alpha \mu^\delta_\beta - \tfrac{1}{2}z\kappa_{\gamma\delta}(\delta^\gamma_\alpha \mu^\delta_\beta + \delta^\gamma_\beta \mu^\delta_\alpha). \tag{9.48}$$

This equation shows that the strain $\eta_{\alpha\beta}$ and hence also the stress $\sigma^{\alpha\beta}$ depends *only* on the deformation of the middle surface and that our kinematic relations, in particular (9.37), will yield zero strain when the shell is subjected to a rigid-body displacement.

## 9.3. Stress Resultants and Equilibrium

In a shell the stress systems of plates and slabs are combined. There is a *membrane force tensor* $N^{\alpha\beta}$ corresponding to the tension-and-shear tensor of the plane slab and a *moment tensor* $M^{\alpha\beta}$, whose components are bending and twisting moments, and, as in a plate, it is inseparable from transverse shear forces $Q^\alpha$. All these forces and moments participate in establishing the equilibrium of the shell element, but because of its curvature they interfere with each other in a more complicated way than in plates and slabs.

Our present task is to define all these stress resultants in terms of the stresses in the shell.

We select an arbitrary line element $d\mathbf{s} = dx^\gamma\, \mathbf{a}_\gamma$ on the middle surface. The total of all the normals to the middle surface emanating from points of $d\mathbf{s}$ forms a section element (Figure 9.3). It is roughly a rectangle of height $h$ and length $ds$, but since the middle surface has curvature and twist, the normals are not precisely parallel to each other and do not even lie in one plane; hence the section element is curved and twisted. We have to keep this in mind when calculating stress resultants.

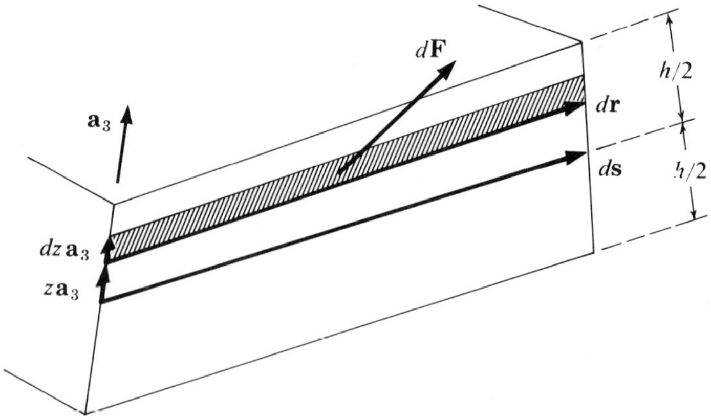


FIGURE 9.3 *Section through a shell.*

Inside the section element we isolate a subelement of height $dz$, as indicated in Figure 9.3. The line element vector $d\mathbf{r}$, which is the base of this element, differs from $d\mathbf{s}$ in magnitude and direction. It has the same components $dx^\gamma$, but in a local reference frame $\mathbf{g}_\gamma$,

$$d\mathbf{r} = dx^\gamma\, \mathbf{g}_\gamma\,. \tag{9.49}$$

The area of the subelement is

$$d\mathbf{A} = d\mathbf{r} \times \mathbf{a}_3\, dz = dx^\gamma\, dz\, \bar{\epsilon}_{\gamma 3\alpha}\mathbf{g}^\alpha = dA_\alpha\, \mathbf{g}^\alpha$$

with the components

$$dA_\alpha = \bar{\epsilon}_{\alpha\gamma}\, dx^\gamma\, dz. \tag{9.50}$$

If we assume that the material of the shell lies behind the area element, the vector $d\mathbf{A}$ has the direction of the outer normal.

Across the subelement a force

$$d\mathbf{F} = dF^\delta\, \mathbf{g}_\delta + dF^3\, \mathbf{g}_3$$

is transmitted. With the definition (4.6) of the stress $\sigma^{ij}$, we can write it in the form

$$d\mathbf{F} = \sigma^{\alpha\delta}\, dA_\alpha\, \mathbf{g}_\delta + \sigma^{\alpha 3}\, dA_\alpha\, \mathbf{g}_3\,. \tag{9.51}$$

The first term represents a force in the tangential plane of the middle surface, while the second one is a force normal to the shell. We deal first with this latter one because it is the simplest. Integrating across the shell thickness $h$, we find the resultant

$$\int dF^3 \, \mathbf{g}_3 = \mathbf{a}_3 \int \sigma^{\alpha 3} \, dA_\alpha = \mathbf{a}_3 \, dx^\gamma \int_{-h/2}^{+h/2} \sigma^{\alpha 3} \, \bar{\epsilon}_{\alpha\gamma} \, dz.$$

The permutation tensor $\bar{\epsilon}_{\alpha\gamma}$ depends on the local metric $g_{\alpha\gamma}$, but we may use (9.22) to express it in terms of $\epsilon_{\alpha\gamma}$, which is defined in terms of $a_{\alpha\gamma}$ and, hence, does not depend on $z$. The resultant force is then

$$\int dF^3 \, \mathbf{g}_3 = \mathbf{a}_3 \, dx^\gamma \, \epsilon_{\alpha\gamma} \int_{-h/2}^{+h/2} \sigma^{\alpha 3} \mu \, dz.$$

We define

$$Q^\alpha = - \int_{-h/2}^{+h/2} \sigma^{\alpha 3} \mu \, dz \tag{9.52}$$

and have then the vertical force

$$\int dF^3 \, \mathbf{g}_3 = - \epsilon_{\alpha\gamma} \, dx^\gamma \, Q^\alpha \, \mathbf{a}_3 \, . \tag{9.53}$$

$Q^\alpha$ is the *transverse shear force* of the shell. Its interpretation is the same as that of the shear force in a plate. Equation (9.52) differs from its counterpart (7.62a) for the plate by the presence of the factor $\mu$, which accounts for the curvature of the shell and all its consequences.

We now turn to the first term in (9.51) and subject it to a similar treatment. This term contains a factor $\mathbf{g}_\delta$ but while $\mathbf{g}_3 = \mathbf{a}_3$ is a constant, $\mathbf{g}_\delta$ depends, through (9.1), on $z$, and we have to make use of this equation to express the force in the reference frame $\mathbf{a}_\beta$:

$$\int dF^\delta \, \mathbf{g}_\delta = \int_{-h/2}^{+h/2} \sigma^{\alpha\delta} \bar{\epsilon}_{\alpha\gamma} \, dx^\gamma \, dz \, \mathbf{g}_\delta = \epsilon_{\alpha\gamma} \, dx^\gamma \, \mathbf{a}_\beta \int_{-h/2}^{+h/2} \sigma^{\alpha\delta} \mu \mu_\delta^\beta \, dz.$$

We define

$$N^{\alpha\beta} = \int_{-h/2}^{+h/2} \sigma^{\alpha\delta} \mu \mu_\delta^\beta \, dz \tag{9.54}$$

and have then

$$\int dF^\delta \, \mathbf{g}_\delta = \epsilon_{\alpha\gamma} dx^\gamma N^{\alpha\beta} \mathbf{a}_\beta \, . \tag{9.55}$$

The force $d\mathbf{F}$ of (9.51) has a lever arm $z\mathbf{a}_3$ with respect to the center of the section and hence a moment

$$d\mathbf{M} = z\mathbf{a}_3 \times d\mathbf{F} = z(\sigma^{\alpha\delta} \mathbf{a}_3 \times \mathbf{g}_\delta + \sigma^{\alpha 3} \mathbf{a}_3 \times \mathbf{g}_3) \, dA_\alpha \, .$$

Because of $\mathbf{a}_3 = \mathbf{g}_3$, the second term in the parentheses equals zero and the total moment of the forces $d\mathbf{F}$ in the section element is

$$\int d\mathbf{M} = \int_{-h/2}^{+h/2} z\sigma^{\alpha\delta}\mathbf{a}_3 \times \mu_\delta^\beta \mathbf{a}_\beta \epsilon_{\alpha\gamma}\mu\, dx^\gamma\, dz = \epsilon_{\alpha\gamma}\epsilon_{\beta\zeta}\mathbf{a}^\zeta\, dx^\gamma \int_{-h/2}^{+h/2} \sigma^{\alpha\delta}\mu_\delta^\beta \mu z\, dz.$$

We define

$$M^{\alpha\beta} = -\int_{-h/2}^{+h/2} \sigma^{\alpha\delta}\mu_\delta^\beta \mu z\, dz \tag{9.56}$$

and have then

$$\int d\mathbf{M} = -\epsilon_{\alpha\gamma}\epsilon_{\beta\zeta}\, dx^\gamma\, M^{\alpha\beta}\mathbf{a}^\zeta. \tag{9.57}$$

The definition of the moment tensor $M^{\alpha\beta}$ is similar to the definition (7.62c) for the plate; the difference lies again in the presence of factors which account for the curvature and the twist of the shell.

The stress resultants $Q^\alpha$, $N^{\alpha\beta}$, $M^{\alpha\beta}$ must satisfy certain differential equations, which express the equilibrium of a shell element. We proceed in the same way as we did for the plate, obtaining the equilibrium conditions by integrating (6.6) without ever seeing a shell element. The only difference lies in the factor by which we multiply this equation. It must be so chosen that the three integrals obtained coincide with those occurring in (9.52), (9.54), and (9.56).

Just as we did on page 128 for the plate, we let $i = \gamma$ in (6.6) and split the three-term sum over $j$ into a two-term sum over $\beta$ plus an extra term for $j = 3$:

$$\sigma^{\gamma\beta}|_\beta + \sigma^{\gamma 3}|_3 + X^\gamma = 0. \tag{9.58}$$

The covariant derivatives in this equation are, of course, the three-dimensional derivatives based on the metric at the point $B$, Figure 9.1; that is, they are based on $g_{\alpha\beta}$ and $g_{33}$. We use (5.23d), (9.18), and (9.20) to express those derivatives in terms of the two-dimensional covariant derivative based on the metric of the middle surface. Beginning with $\sigma^{\gamma\beta}|_\beta$, we find

$$\begin{aligned}\sigma^{\gamma\beta}|_\beta &= \sigma^{\gamma\beta}{}_{,\beta} + \sigma^{\zeta\beta}\bar{\Gamma}^\gamma_{\zeta\beta} + \sigma^{\gamma\zeta}\bar{\Gamma}^\beta_{\beta\zeta} + \sigma^{3\beta}\bar{\Gamma}^\gamma_{3\beta} + \sigma^{\gamma 3}\bar{\Gamma}^\beta_{\beta 3}\\ &= \sigma^{\gamma\beta}{}_{,\beta} + \sigma^{\zeta\beta}(\Gamma^\gamma_{\zeta\beta} + \lambda^\gamma_\delta \mu^\delta_\zeta\|_\beta) + \sigma^{\gamma\zeta}(\Gamma^\beta_{\beta\zeta} + \lambda^\beta_\delta \mu^\delta_\beta\|_\zeta)\\ &\quad - \sigma^{3\beta}\lambda^\gamma_\delta b^\delta_\beta - \sigma^{\gamma 3}\lambda^\beta_\delta b^\delta_\beta.\end{aligned}$$

We may now use (8.38) to combine three of the terms into $\sigma^{\gamma\beta}\|_\beta$, and we have

$$\sigma^{\gamma\beta}|_\beta = \sigma^{\gamma\beta}\|_\beta + \sigma^{\zeta\beta}\lambda^\gamma_\delta \mu^\delta_\zeta\|_\beta + \sigma^{\gamma\zeta}\lambda^\beta_\delta \mu^\delta_\beta\|_\zeta - (\sigma^{3\beta}\lambda^\gamma_\delta + \sigma^{\gamma 3}\lambda^\beta_\delta)b^\delta_\beta.$$

Since we later shall multiply (9.58) by $\mu^\alpha_\gamma \mu$, we now apply this factor to $\sigma^{\gamma\beta}|_\beta$ and make use of (9.4) and (9.16) to find

$$\begin{aligned}\sigma^{\gamma\beta}|_\beta\,\mu_\gamma^\alpha\mu &= \sigma^{\gamma\beta}\|_\beta\,\mu_\gamma^\alpha\mu + \sigma^{\gamma\beta}\mu_\gamma^\alpha\|_\beta\,\mu + \sigma^{\gamma\zeta}\mu_\gamma^\alpha\,\mu\|_\zeta - (\sigma^{3\beta}\,\delta_\delta^\alpha + \sigma^{\gamma 3}\lambda_\delta^\beta\,\mu_\gamma^\alpha)\mu b_\beta^\delta \\ &= (\sigma^{\gamma\beta}\mu_\gamma^\alpha\mu)\|_\beta - (\sigma^{3\beta}b_\beta^\alpha + \sigma^{\gamma 3}\lambda_\delta^\beta\,\mu_\gamma^\alpha\,b_\beta^\delta)\mu.\end{aligned}$$

When $\sigma^{\gamma 3}|_3$ is processed in the same manner, many of the Christoffel symbols turn out to be zero and we have

$$\sigma^{\gamma 3}|_3 = \sigma^{\gamma 3}{}_{,3} + \sigma^{\zeta 3}\Gamma_{3\zeta}^\gamma = \sigma^{\gamma 3}{}_{,3} - \sigma^{\zeta 3}\lambda_\delta^\gamma\,b_\zeta^\delta$$

and, after multiplication by $\mu_\gamma^\alpha\,\mu$ and use of (9.2),

$$\sigma^{\gamma 3}|_3\,\mu_\gamma^\alpha\,\mu = (\sigma^{\gamma 3}{}_{,3}\,\mu_\gamma^\alpha - \sigma^{\zeta 3}b_\zeta^\alpha)\mu = (\sigma^{\gamma 3}{}_{,3}\,\mu_\gamma^\alpha + \sigma^{\zeta 3}\mu_{\zeta,3}^\alpha)\mu = (\sigma^{\gamma 3}\mu_\gamma^\alpha)_{,3}\mu.$$

We now return to (9.58), multiply it by $\mu_\gamma^\alpha\,\mu\,dz$, introduce the results just obtained, and integrate across the shell thickness $h$:

$$\int_{-h/2}^{+h/2}(\sigma^{\gamma\beta}\mu_\gamma^\alpha\mu)\|_\beta\,dz - b_\beta^\alpha\int_{-h/2}^{+h/2}\sigma^{3\beta}\mu\,dz \\ + \int_{-h/2}^{+h/2}[-\sigma^{\gamma 3}\mu_\gamma^\alpha\,\lambda_\delta^\beta\,b_\beta^\delta + (\sigma^{\gamma 3}\mu_\gamma^\alpha)_{,3}]\mu\,dz + \int_{-h/2}^{+h/2}X^\gamma\mu_\gamma^\alpha\,\mu\,dz = 0. \quad (9.59)$$

In the first two terms we recognize $N^{\beta\alpha}\|_\beta$ and $b_\beta^\alpha\,Q^\beta$ from (9.54) and (9.52) and to the third integral we apply (9.17), which reduces it to

$$\int_{-h/2}^{+h/2}[\sigma^{\gamma 3}\mu_\gamma^\alpha\,\mu_{,3} + (\sigma^{\gamma 3}\mu_\gamma^\alpha)_{,3}\,\mu]\,dz = \Big[\sigma^{\gamma 3}\mu_\gamma^\alpha\,\mu\Big]_{-h/2}^{+h/2}. \quad (9.60)$$

This is the difference of the surface tractions acting on the faces $z = \pm h/2$, each of them multiplied by a factor, which reduces the local metric to that of the middle surface. Similarly, the last integral represents the resultant of the local volume forces, and both terms together can be combined into the tangential load component

$$p^\alpha = \Big[\sigma^{\gamma 3}\mu_\gamma^\alpha\,\mu\Big]_{-h/2}^{+h/2} + \int_{-h/2}^{+h/2}X^\gamma\mu_\gamma^\alpha\,\mu\,dz.$$

The equilibrium condition appears then in its final form:

$$N^{\beta\alpha}\|_\beta + Q^\beta b_\beta^\alpha + p^\alpha = 0. \quad (9.61)$$

It assures the equilibrium of forces in the direction $\mathbf{a}_\alpha$ ($\alpha = 1, 2$).

To obtain the condition of equilibrium of forces normal to the shell, we return to (6.6) and now let $i = 3$ and $j = \alpha, 3$:

$$\sigma^{3\alpha}|_\alpha + \sigma^{33}|_3 + X^3 = 0. \quad (9.62)$$

When we apply (5.23d) to the first term, we recognize that one of the Christoffel symbols is zero, and we use equations (9.18) through (9.20) to reduce the other ones to terms defined on the middle surface:

$$\begin{aligned}\sigma^{3\alpha}|_\alpha &= \sigma^{3\alpha}{}_{,\alpha} + \sigma^{\gamma\alpha}\Gamma_{\gamma\alpha}^3 + \sigma^{3\gamma}\Gamma_{\gamma\alpha}^\alpha + \sigma^{3\alpha}\Gamma_{3\alpha}^3 + \sigma^{33}\Gamma_{3\alpha}^\alpha \\ &= \sigma^{3\alpha}{}_{,\alpha} + \sigma^{\gamma\alpha}\mu_\gamma^\beta\,b_{\alpha\beta} + \sigma^{3\gamma}(\Gamma_{\gamma\alpha}^\alpha + \lambda_\delta^\alpha\,\mu_\alpha^\delta\|_\gamma) - \sigma^{33}\lambda_\delta^\alpha\,b_\alpha^\delta.\end{aligned}$$

While $\sigma^{\alpha\beta}$ is a two-dimensional stress tensor, $\sigma^{3\alpha}$ has only one free superscript and is a vector. Therefore, its two-dimensional covariant derivative follows (8.32). Its two terms can easily be recognized in the last expression, and after multiplication by $\mu$, we have

$$\sigma^{3\alpha}|_\alpha \mu = \sigma^{\gamma\alpha}\mu^\beta_\gamma \mu b_{\alpha\beta} + \sigma^{3\alpha}\|_\alpha \mu + \sigma^{3\gamma}\mu\|_\gamma + \sigma^{33}\mu_{,3}\,.$$

In the last two terms of this equation, use has been made of (9.16) and (9.17).

It is easily seen that $\sigma^{33}|_3 = \sigma^{33}{}_{,3}$, and when we now multiply (9.62) by $\mu$ and integrate across the shell thickness, we arrive at the following equation:

$$b_{\alpha\beta}\int_{-h/2}^{+h/2}\sigma^{\gamma\alpha}\mu^\beta_\gamma \mu\, dz + \int_{-h/2}^{+h/2}(\sigma^{3\alpha}\|_\alpha \mu + \sigma^{3\alpha}\mu\|_\alpha)\, dz$$
$$+ \int_{-h/2}^{+h/2}(\sigma^{33}\mu_{,3} + \sigma^{33}{}_{,3}\,\mu)\, dz + \int_{-h/2}^{+h/2} X^3 \mu\, dz = 0.$$

The first integral is $N^{\alpha\beta}$ from (9.54); the second one is the $\alpha$-derivative of $Q^\alpha$ from (9.52), and the third one can be integrated and combined with the last one to yield

$$\Big[\sigma^{33}\mu\Big]_{-h/2}^{+h/2} + \int_{-h/2}^{+h/2} X^3\mu\, dz = p^3,$$

that is, the normal component of the distributed load in the direction of positive $z$. The entire equation then assumes the following, final form:

$$N^{\alpha\beta}b_{\alpha\beta} - Q^\alpha\|_\alpha + p^3 = 0, \tag{9.63}$$

and this is the equilibrium condition for forces normal to the shell.

To insure the moment equilibrium of a shell element, we return to (9.59) and insert a factor $z$ under each integral. Since the sum of the integrands in this equation vanishes, this is a permissible operation, exactly as the multiplication of the original equilibrium condition by $\mu^\alpha_\gamma \mu$, which is already incorporated in (9.59). In the modified equation, the first integral is the covariant derivative of $M^{\beta\alpha}$ from (9.56); in the third integral we repeat the operation which led to (9.60) and integrate by parts:

$$\int_{-h/2}^{+h/2}[\sigma^{\gamma 3}\mu^\alpha_\gamma \mu_{,3} + (\sigma^{\gamma 3}\mu^\alpha_\gamma)_{,3}\,\mu]z\, dz = \int_{-h/2}^{+h/2}(\sigma^{\gamma 3}\mu^\alpha_\gamma \mu)_{,3}\, z\, dz$$
$$= \Big[\sigma^{\gamma 3}\mu^\alpha_\gamma \mu z\Big]_{-h/2}^{+h/2} - \int_{-h/2}^{+h/2}\sigma^{\gamma 3}\mu^\alpha_\gamma \mu\, dz.$$

We can now write the modified equation (9.59) in the following form:

$$-M^{\beta\alpha}\|_\beta - b^\alpha_\beta \int_{-h/2}^{+h/2}\sigma^{3\beta}\mu z\, dz + \Big[\sigma^{\gamma 3}\mu^\alpha_\gamma \mu z\Big]_{-h/2}^{+h/2} - \int_{-h/2}^{+h/2}\sigma^{\beta 3}\mu^\alpha_\beta \mu\, dz$$
$$+ \int_{-h/2}^{+h/2} X^\gamma \mu^\alpha_\gamma \mu z\, dz = 0.$$

Making use of (9.2), we may combine the first two of the integrals:

$$-\int_{-h/2}^{+h/2} \sigma^{3\beta}(z b_\beta^\alpha + \mu_\beta^\alpha)\mu \, dz = -\int_{-h/2}^{+h/2} \sigma^{3\beta}\, \delta_\beta^\alpha \, \mu \, dz = Q^\alpha ;$$

and the third and fifth terms are the contributions of the surface tractions and the body forces to a moment load, which we denote by $-m^\alpha$. Thus we arrive at the final form of our equation:

$$M^{\beta\alpha}\|_\beta - Q^\alpha + m^\alpha = 0. \tag{9.64}$$

This equation, which stands for two component equations, assures the moment equilibrium with respect to all axes tangent to the middle surface.

In deriving (9.61), (9.63), and (9.64), we started from (6.6), which is a condition imposed upon the derivatives of the stresses in order to insure equilibrium of a volume element. Like (6.6), the three shell equations derived from it are differential equations and are essential conditions which the stress resultants must satisfy. There is still another equilibrium condition, the one for the moments about a normal to the shell, and this equation has a quite different position within the set of shell equations, as we shall presently see.

We start from (4.8), let $i, j = \alpha, \beta$ and make use of the permutation tensor, writing

$$\bar{\epsilon}_{\alpha\beta}\, \sigma^{\alpha\beta} = 0. \tag{9.65}$$

We apply (9.21), multiply by $\mu$ and integrate:

$$\epsilon_{\gamma\delta} \int_{-h/2}^{+h/2} \sigma^{\alpha\beta} \mu_\alpha^\gamma \mu_\beta^\delta \, \mu \, dz = 0.$$

We split the factor $\mu_\beta^\delta$ according to (9.2) and have

$$\epsilon_{\gamma\delta} \int_{-h/2}^{+h/2} \sigma^{\alpha\beta} \mu_\alpha^\gamma \, \delta_\beta^\delta \, \mu \, dz - \epsilon_{\gamma\delta}\, b_\beta^\delta \int_{-h/2}^{+h/2} \sigma^{\alpha\beta} \mu_\alpha^\gamma \, \mu z \, dz = 0,$$

whence

$$\epsilon_{\gamma\delta}(N^{\delta\gamma} + b_\beta^\delta M^{\beta\gamma}) = 0$$

or, after some changes in the notation for the dummy indices,

$$\epsilon_{\alpha\beta}(N^{\alpha\beta} + b_\gamma^\alpha M^{\gamma\beta}) = 0. \tag{9.66}$$

This is the desired equilibrium condition. Different from the other ones, it is algebraic and therefore does not control the variation of the stress resultants from one point to the next, but is a condition imposed on their values at any single point. It flows immediately from the statement that the stress tensor is symmetric and is nothing else but this statement, expressed in stress resultants. This symmetry, which through Hooke's law corresponds to the symmetry of the strain tensor, is part of our fundamental concepts and not an additional condition imposed upon the stresses and the stress resultants. This is the reason why (9.66) is not part of the fundamental shell equations, but is a surplus equation, which the solution of the fundamental equations will satisfy automatically. This does not mean that the equation is entirely worthless. There have been many attempts at finding shortcuts to avoid the complexity of the shell equations, and many of these attempts have introduced simplifying assumptions leading to a violation of (9.66). This is not necessarily a grievous deficiency of such attempts, since the violation *may* be very slight, but in any simplified theory (9.66) deserves careful examination.

Equations (9.61), (9.63), and (9.64) are the essential equilibrium conditions of the shell. It is possible to use (9.64) to eliminate the transverse shear force $Q^\alpha$ from the other two. The result of the simple calculation is the following pair of equations:

$$\left.\begin{aligned} N^{\beta\alpha}\|_\beta + M^{\beta\gamma}\|_\beta b^\alpha_\gamma &= -p^\alpha - m^\beta b^\alpha_\beta, \\ N^{\alpha\beta} b_{\alpha\beta} - M^{\alpha\beta}\|_{\alpha\beta} &= -p^3 + m^\alpha\|_\alpha . \end{aligned}\right\} \tag{9.67}$$

When using these equations, one should keep in mind that the tensors $N^{\alpha\beta}$ and $M^{\alpha\beta}$ are not symmetric and that in the second derivative the order of the covariant differentiations cannot be interchanged. Together they represent three component equations, associated with the equilibrium of forces in tangential and normal directions.

When a shell is thin enough, it seems plausible that the moments $M^{\alpha\beta}$ cannot make a substantial contribution to the equilibrium. Practical experience with solved problems shows that this is true if the boundary conditions are favorable and if there are no discontinuities in the curvature of the shell and the distribution of the load. Naturally, moment loads $m^\alpha$ must be excluded in this case. A shell theory based upon the assumption that $M^{\alpha\beta} \equiv 0$ is called *membrane theory*.

If the assumption is accepted, it follows immediately from (9.64) that also $Q^\alpha = 0$, and then (9.61) and (9.63) as well as the pair (9.67) reduce to

$$N^{\beta\alpha}\|_\beta = -p^\alpha, \qquad N^{\alpha\beta} b_{\alpha\beta} = -p^3. \tag{9.68a, b}$$

In the absence of all moments (9.66) simply reads $\epsilon_{\alpha\beta} N^{\alpha\beta} = 0$ and shows that $N^{12} = N^{21}$. The two component equations (9.68a) and the single equation (9.68b) then contain only three unknowns $N^{11}$, $N^{12}$, $N^{22}$ and can be solved for them if there are sufficiently many boundary conditions in terms of the stress resultants to make the solution unique. A structure of this kind is called statically determinate. If desired, its deformation can be calculated after the stress system has been found.

## 9.4. Elastic Law

After having established the kinematic relations (9.38) and (9.45) and the equilibrium conditions (9.61), (9.63), and (9.64), we have only one step left to complete the fundamental set of shell equations. We still must write the elastic law, which connects the stress resultants $N^{\alpha\beta}$ and $M^{\alpha\beta}$ with the midsurface strains $\varepsilon_{\alpha\beta}$ and $\kappa_{\alpha\beta}$.

The procedure is simple. We start from the definitions (9.54) and (9.56) of the stress resultants, use Hooke's law (7.29) to express the stress $\sigma^{\alpha\delta}$ in terms of the local strain $\varepsilon_\gamma^\alpha$, which we now denote by $\eta_\gamma^\alpha$, and use the kinematic equation (9.48) to express $\eta_\gamma^\alpha$ in terms of the midsurface strains. Equations (9.54) and (9.56) contain contravariant stresses while Hooke's law, in the form we want to use, has components of mixed variance and (9.48) is written in covariant components. This makes it necessary first to correlate all these components, using the metric tensor $g^{\alpha\beta}$ and (9.6b).

We write for the contravariant stress components

$$\sigma^{\alpha\delta} = \sigma_\gamma^\alpha g^{\gamma\delta} = \sigma_\gamma^\alpha \lambda_\rho^\gamma \lambda_\sigma^\delta a^{\rho\sigma}$$

and use the plane stress form (7.29) of Hooke's law:

$$\sigma_\gamma^\alpha = \frac{E}{1-\nu^2}[(1-\nu)\eta_\gamma^\alpha + \nu\delta_\gamma^\alpha \eta_\zeta^\zeta]$$

to obtain

$$\sigma^{\alpha\delta} = \frac{E}{1-\nu^2}[(1-\nu)\eta_\gamma^\alpha + \nu\delta_\gamma^\alpha \eta_\zeta^\zeta]\lambda_\rho^\gamma \lambda_\sigma^\delta a^{\rho\sigma}.$$

This may be introduced into (9.54):

$$\begin{aligned} N^{\alpha\beta} &= \frac{E}{1-\nu^2} a^{\rho\sigma} \int_{-h/2}^{+h/2} [(1-\nu)\eta_\gamma^\alpha + \nu\delta_\gamma^\alpha \eta_\zeta^\zeta]\lambda_\rho^\gamma \lambda_\sigma^\delta \mu\mu_\delta^\beta \, dz \\ &= \frac{E}{1-\nu^2} a^{\rho\beta} \int_{-h/2}^{+h/2} [(1-\nu)\eta_\gamma^\alpha + \nu\delta_\gamma^\alpha \eta_\zeta^\zeta]\lambda_\rho^\gamma \mu \, dz. \end{aligned}$$

The strains $\eta_\gamma^\alpha$, $\eta_\zeta^\zeta$ must now be expressed in terms of the covariant components $\eta_{\zeta\delta}$:

$$\begin{aligned} \eta_\gamma^\alpha &= \eta_{\gamma\delta} g^{\alpha\delta} = \eta_{\gamma\delta} \lambda_\mu^\alpha \lambda_\nu^\delta a^{\mu\nu} = \eta_{\zeta\delta}\delta_\gamma^\zeta \lambda_\mu^\alpha \lambda_\nu^\delta a^{\mu\nu}, \\ \eta_\zeta^\zeta &= \eta_{\zeta\delta} \lambda_\mu^\zeta \lambda_\nu^\delta a^{\mu\nu}, \end{aligned}$$

whence

$$N^{\alpha\beta} = \frac{E}{1-\nu^2} a^{\rho\beta} a^{\mu\nu} \int_{-h/2}^{+h/2} [(1-\nu)\delta_\gamma^\zeta \lambda_\mu^\alpha + \nu\delta_\gamma^\alpha \lambda_\mu^\zeta]\eta_{\zeta\delta} \lambda_\nu^\delta \lambda_\rho^\gamma \mu \, dz.$$

Now it is time to make use of (9.48), which we rewrite with appropriate changes of the indices:

$$\eta_{\zeta\delta} = \varepsilon_{\sigma\tau}\,\mu_\zeta^\sigma \mu_\delta^\tau - \tfrac{1}{2} z\kappa_{\sigma\tau}(\delta_\zeta^\sigma \mu_\delta^\tau + \delta_\delta^\sigma \mu_\zeta^\tau).$$

Introducing this into the preceding equation yields the following expression for $N^{\alpha\beta}$:

$$N^{\alpha\beta} = \frac{E}{1-\nu^2}\, a^{\rho\beta} a^{\mu\nu} \int_{-h/2}^{+h/2} [(1-\nu)\delta_\gamma^\zeta \lambda_\mu^\alpha + \nu\delta_\gamma^\alpha \lambda_\mu^\zeta]$$

$$\times\, [\varepsilon_{\sigma\nu}\,\mu_\zeta^\sigma - \tfrac{1}{2} z(\kappa_{\zeta\nu} + \kappa_{\delta\tau}\,\mu_\zeta^\tau \lambda_\nu^\delta)]\lambda_\rho^\gamma\,\mu\, dz. \quad (9.69)$$

The integrand in this equation contains the constants $\varepsilon_{\sigma\nu}$, $\kappa_{\zeta\nu}$, $\kappa_{\delta\tau}$, an explicit factor $z$ in the second bracket, and quantities $\mu_\beta^\alpha$, $\lambda_\beta^\alpha$, $\mu$, which depend on $z$ and the curvature tensor according to (9.2), (9.5), and (9.12). When they are explicitly expressed in powers of $z$, the integrand becomes a power series and the integration can be performed separately for each term. The definite integrals of the odd powers are zero, and the integrals of the even powers lead us to introduce the extensional stiffness

$$D = \frac{Eh}{1-\nu^2} \quad (9.70a)$$

and the bending stiffness

$$K = \frac{Eh^3}{12(1-\nu^2)}. \quad (9.70b)$$

We decide to drop terms with $h^5$ and higher powers of $h$. This decision is made in addition to the assumption of the conservation of normals, but it is not independent of it. Since the conservation of normals is only approximately true, it is worthless to carry the higher powers of $h$, and even the terms in $h^3$ are not exact. They are carried, however, since this is the only way one can obtain a formulation of the shell equations which does not violate one or another of the basic laws of mechanics.

Working out the details is lengthy, but simple, tensor algebra. The result is the following relation:

$$N^{\alpha\beta} = D[(1-\nu)\varepsilon^{\alpha\beta} + \nu\varepsilon_\zeta^\zeta a^{\alpha\beta}]$$

$$- K\Bigg[\frac{1-\nu}{2}[2a^{\beta\delta}b^{\alpha\gamma} + a^{\beta\gamma}b^{\alpha\delta} + a^{\alpha\delta}b^{\beta\gamma} - b_\lambda^\lambda(a^{\alpha\delta}a^{\beta\gamma} + a^{\alpha\gamma}a^{\beta\delta})]$$

$$+ \nu(a^{\alpha\beta}b^{\gamma\delta} + a^{\gamma\delta}b^{\alpha\beta} - a^{\alpha\beta}a^{\gamma\delta}b_\lambda^\lambda)\Bigg]\kappa_{\gamma\delta}. \quad (9.71)$$

The only difference between (9.54) and (9.56) lies in a factor $z$ under the integral, which is present in $M^{\alpha\beta}$ and is missing in $N^{\alpha\beta}$. We may, therefore, obtain a formula for the moments by retracing all the steps leading up to (9.69), just inserting this factor $z$. Then again when the integrand is written as a polynomial in $z$, there will be no term without $z$ and, within the limits of accuracy chosen, only the term with $z^2$ makes a contribution. Working out the algebra yields

$$M^{\alpha\beta} = K[(1-\nu)(b^{\alpha}_{\gamma}\varepsilon^{\gamma\beta} - b^{\lambda}_{\lambda}\varepsilon^{\alpha\beta}) + \nu(b^{\alpha\beta} - b^{\lambda}_{\lambda}a^{\alpha\beta})\varepsilon^{\mu}_{\mu} - \tfrac{1}{2}(1-\nu)(\kappa^{\alpha\beta} + \kappa^{\beta\alpha}) + \nu a^{\alpha\beta}\kappa^{\lambda}_{\lambda}]. \quad (9.72)$$

with

$$\kappa^{\alpha\beta} = \kappa^{\beta}_{\gamma}a^{\alpha\gamma} = \kappa_{\gamma\delta}a^{\beta\delta}a^{\alpha\gamma}. \quad (9.73)$$

The two equations (9.71) and (9.72) represent the elastic law of the shell. Together with the kinematic relations (9.38), (9.45) and the equilibrium conditions (9.61), (9.63), (9.64) they are the fundamental equations of shell theory.

## Problems

9.1. Derive the relation

$$\lambda^{\alpha}_{\beta}\mu^{\delta}_{\alpha} = \delta^{\delta}_{\beta},$$

which is a counterpart of (9.4).

9.2. Calculate $\hat{\mathbf{g}}_{\alpha} = \mathbf{r}_{,\alpha}$ and thence $\hat{g}_{\alpha\beta} = \hat{\mathbf{g}}_{\alpha}\cdot\hat{\mathbf{g}}_{\beta}$ and $2_{\eta\alpha\beta} = \hat{g}_{\alpha\beta} - g_{\alpha\beta}$. Compare the result with (9.37).

9.3. Apply the fundamental equations of shell theory to a spherical shell of radius $a$. As coordinates, use the colatitude $\phi = x^1$ measured along a meridian from the apex and the longitude $\theta = x^2$.

## References

For two generations, the last chapters of Love's book [19] have been the basic reference in matters of shell theory; and the book still has a place of honor in the field. A larger supply of solved problems is found in newer books, among which [10] and [11] are exclusively devoted to shells, while [12] and [34] also include plates. All these books use component equations in coordinates adapted to special classes of shell geometry. Love [19] and Novozhilov [25] give general equations based on coordinates following the lines of curvature. Green–Zerna [13] present a general shell theory in tensor form.

There are several ways of deriving shell equations and each of them has its protagonists. A large amount of literature has sprung up using one or the other and comparing the results. The approach used in this book, based upon strict adherence to the postulate of conservation of normals, is usually attributed to Kirchhoff and Love. It seems that the first equations which, by their standards, are exact, are the equations for the circular cylinder given in [8] and reproduced in [10] and [11]. The subject has since then been taken up by many authors. Equations for a general shell of revolution may be found in [10].

A very general presentation of shell theory, using tensor notation and general curvilinear coordinates, has been given by Naghdi [24] and Marlowe [21]. The presentation in this book has made much use of these, but goes beyond them, drawing a distinct line between strain quantities and displacement quantities (including the so-called displacement gradients), admitting only strain in the constitutive equations.

CHAPTER 10

# Elastic Stability

WE DEFINE AS a *structure* a solid body (or even a number of connected bodies) which is so supported that it can carry loads. In general, elastic structures are deterministic. To each (positive or negative) increment of load they respond with a definite change of stress and deformation. This corresponds to the fact that the displacement vector $u_k$ is determined by a linear differential equation (6.8) which, together with the appropriate boundary conditions, has a unique solution.

Observation shows that there are exceptions to the deterministic behavior. There are elastic structures which, under a certain *critical load*, can be in equilibrium in several adjacent states of stress and the corresponding states of deformation. Since all these stress states pertain to the same load, the elastic structure is in neutral equilibrium and a spontaneous transition from one state of stress to another is possible. Such a transition is known as *buckling* of the structure. The buckling of a straight column under its Euler load is an example.

It is obvious that this phenomenon cannot be described by (6.8). The differential equation which admits to more than one solution must contain some sort of nonlinearity. To find it, we inspect the sources of the linearity of (6.8), from which we shall now drop the dynamic term $2\rho\ddot{u}^i$. This equation summarizes Hooke's law (4.11), the kinematic relation (6.1), and the equilibrium condition (6.6). Hooke's law postulates that the material has linear elastic behavior and we do not intend considering a different material. The kinematic relation (6.1) has been obtained by linearizing the exact equation (6.2). This linearization is possible as long as $u_i|_j \ll 1$. In a sequence of adjacent states of deformation this is true for all if it is true for one. Since we want to study a neutral equilibrium within the range of small deformation, we have no reason not to use the linearized form. More subtle and most important is the linearization of the equilibrium condition (6.6). The element shown in Figure 6.2 is the undeformed element, and (6.6) formulates the equilibrium of forces acting on this element, while in reality they are acting on a deformed element. Since elastic deformations are small, this approximation is good enough in most cases, but not here, as we may easily understand.

When the load is gradually increased until it reaches its critical value, the stresses and displacements approach everywhere definite limiting values, which we denote by $\bar{\sigma}^{ij}$ and $\bar{u}_k$. We call this the critical state. As soon as this critical state has been reached, an adjacent state becomes possible, which we call the buckled state and in which the stresses are $\hat{\sigma}^{ij} = \bar{\sigma}^{ij} + \sigma^{ij}$ and the displacements $\hat{u}_k = \bar{u}_k + u_k$. Thus, $\sigma^{ij}$ and $u_k$ are the small increments of stress and displacement which occur during buckling. They are infinitesimal in the sense that if we let the buckled state approach the unbuckled one, then $\sigma^{ij} \to 0$ and $u_k \to 0$.

The equilibrium condition for the basic state contains the basic stress $\bar{\sigma}^{ij}$. To write the condition for the buckled state, it is not enough simply to replace $\bar{\sigma}^{ij}$ by $\hat{\sigma}^{ij}$, because these stresses act in slightly different directions. These differences are proportional to the difference in deformation, i.e. to the buckling displacement $u_k$ and give rise to terms $(\bar{\sigma}^{ij} + \sigma^{ij})u_k$. While $\sigma^{ij}u_k$ is a product of two infinitesimal quantities and has to be dropped, the product $\bar{\sigma}^{ij}u_k$ is of the same order of magnitude as the linear terms in $\sigma^{ij}$ and must be kept. Terms of this type are characteristic for all problems of elastic buckling. They are nonlinear because they are products of a stress and a displacement, but they are still linear if we are looking only for the infinitesimal quantities $\sigma^{ij}$ and $u_k$, whose existence signals the presence of an adjacent equilibrium. Thus we arrive at the remarkable situation where we are dealing with a differential equation which has nonunique solutions, but where the mathematical formalism stays within the domain of linear differential equations.

After these preparations we may formulate the differential equations of our problem. As before, we use a coordinate system $x^i$ which is deformed with the body. In the critical state (with stresses $\bar{\sigma}^{ij}$ and displacements $\bar{u}_i$) the base vectors are $\bar{\mathbf{g}}_i = \bar{\mathbf{r}}_{,i}$ and the metric tensor is $\bar{g}_{ij}$. This reference frame will be used for writing the equations, and covariant derivatives are understood to use the $\bar{\Gamma}^i_{jk}$ derived from its metric $\bar{g}_{ij}$.

The rectangular body element shown in Figure 10.1 is cut from the body in the critical state. The face $d\mathbf{s} \times d\mathbf{t}$ on its right-hand side has the area

$$dA_l = ds^j \, dt^k \, \bar{\epsilon}_{jkl}$$

and on it the force $\bar{\sigma}^{lm}\bar{\mathbf{g}}_m \, dA_l$ is acting.

In the buckled state the base vectors are

$$\hat{\mathbf{g}}_m = (\bar{\mathbf{r}} + \mathbf{u})_{,m} = \bar{\mathbf{g}}_m + \mathbf{u}_{,m} = \bar{\mathbf{g}}_m + u^n|_m \bar{\mathbf{g}}_n$$

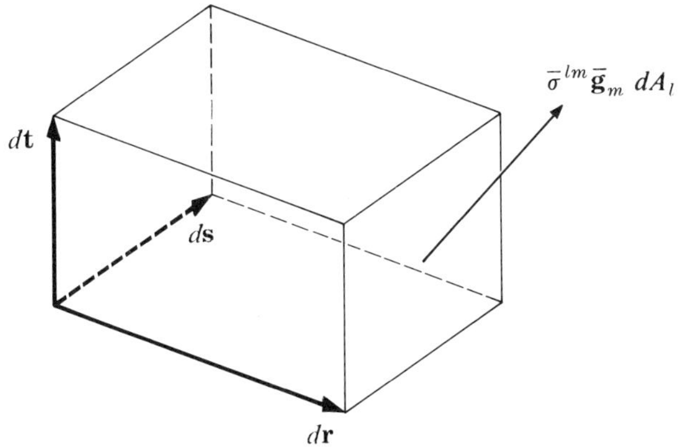


FIGURE 10.1 *Stresses acting on a volume element in the prebuckling state.*

and we write the force then acting as

$$\hat{\sigma}^{lm}\hat{\mathbf{g}}_m\, dA_l;$$

that is, we still refer it to the area before buckling, but we resolve it in components with respect to the postbuckling base vectors $\hat{\mathbf{g}}_m$. We define the incremental stress $\sigma^{lm}$ through the equation

$$\hat{\sigma}^{lm} = \bar{\sigma}^{lm} + \sigma^{lm}.$$

The force on the right-hand face of the volume element may then be written as

$$\hat{\sigma}^{lm}\hat{\mathbf{g}}_m\, dA_l = (\bar{\sigma}^{lm} + \sigma^{lm})(\bar{\mathbf{g}}_m + u^n|_m\,\bar{\mathbf{g}}_n)\, dA_l = (\bar{\sigma}^{lm} + \sigma^{lm} + \bar{\sigma}^{ln}u^m|_n)\bar{\mathbf{g}}_m\, dA_l,$$

omitting from the last member a term which is of second order in the incremental quantities $\sigma^{lm}$ and $u^n$.

We now go back to the derivation of the equilibrium condition (6.6) for the volume element shown in Figure 10.1. We may apply that equation to the stresses before buckling:

$$\bar{\sigma}^{lm}|_l + \bar{X}^m = 0, \tag{10.1}$$

denoting by $\bar{X}^m$ the volume force then acting. After buckling, the forces acting on the pair of faces $d\mathbf{s} \times d\mathbf{t}$ of the body element differ by

$$(\hat{\sigma}^{lm}\hat{\mathbf{g}}_m)_{,i}\, dr^i\, dA_l = (\bar{\sigma}^{lm} + \sigma^{lm} + \bar{\sigma}^{ln}u^m|_n)|_i\,\bar{\mathbf{g}}_m\, dr^i\, ds^j\, dt^k\,\bar{\epsilon}_{ljk}.$$

We add the contributions of the other sides of the body element and have the resultant force of all stresses:

$$[(\bar{\sigma}^{lm} + \sigma^{lm} + \bar{\sigma}^{ln}u^m|_n)|_i\,\bar{\epsilon}_{ljk} + (\;)|_j\,\bar{\epsilon}_{ilk} + (\;)|_k\,\bar{\epsilon}_{ijl}]\bar{\mathbf{g}}_m\, dr^i\, ds^j\, dt^k.$$

According to the argument presented on page 79, this is the same as

$$(\bar{\sigma}^{lm} + \sigma^{lm} + \bar{\sigma}^{ln}u^m|_n)|_l\, \bar{\epsilon}_{ijk}\, dr^i\, ds^j\, dt^k\, \bar{\mathbf{g}}_m$$
$$= [\bar{\sigma}^{lm}|_l + \sigma^{lm}|_l + (\bar{\sigma}^{ln}u^m|_n)|_l]\bar{\epsilon}_{ijk}\, dr^i\, ds^j\, dt^k \bar{\mathbf{g}}_m. \quad (10.2)$$

This force must be in equilibrium with the body force after buckling (see p. 88), which we write in the form

$$(\bar{X}^m + X^m)\bar{\mathbf{g}}_m\, \bar{\epsilon}_{ijk}\, dr^i\, ds^j\, dt^k; \quad (10.3)$$

that is, we refer also the incremental force $X^m$ to the base vectors $\bar{\mathbf{g}}_m$ and to the prebuckling volume of the body element. The sum of the expressions (10.2) and (10.3) must vanish. When we subtract (10.1) from this statement, we are left with

$$[\sigma^{lm}|_l + (\bar{\sigma}^{ln}u^m|_n)|_l + X^m]\bar{\epsilon}_{ijk}\, dr^i\, ds^j\, dt^k \bar{\mathbf{g}}_m = 0.$$

We may drop the scalar factor $\bar{\epsilon}_{ijk}\, dr^i\, ds^j\, dt^k$ and, since the remaining vector has to be zero, each of its components must vanish. After a change of dummy indices this leads to the following form of the equilibrium condition for the buckled state:

$$\sigma^{ji}|_j + (\bar{\sigma}^{jk}\, u^i|_k)|_j + X^i = 0. \quad (10.4)$$

This equation, the elastic law (4.11) or (4.25), and the kinematic relation (6.1), applied to the incremental quantities $\sigma^{ij}$, $\varepsilon_{ij}$, and $u_i$, are the basic equations of the stability problem.

Gravity forces do not change in magnitude or direction when the structure buckles. If they appear as loads, the incremental load $X^i \equiv 0$. Other body forces, like centrifugal forces, may change, but then the increments $X^i$ are proportional to the buckling deformation and thus are part of the unknowns of the problem. Equations (10.4), (4.11), and (6.1) are then linear, homogeneous equations and the structure is in neutral equilibrium when they have a nontrivial solution. This amounts to an eigenvalue problem, the eigenvalue being hidden in the prebuckling stresses $\bar{\sigma}^{ij}$ occurring as coefficients in (10.4).

We shall now apply (10.4) to the buckling of a plane plate. The plate is loaded by edge forces only (hence $X^i \equiv 0$), which act in its middle plane. The stress system is that of a plane slab with the stresses $\bar{\sigma}^{\alpha\beta}$ uniformly distributed across the thickness ($\bar{\sigma}^{\alpha\beta}|_3 \equiv 0$), while $\bar{\sigma}^{\alpha 3} = \bar{\sigma}^{33} = 0$. We introduce the tension-and-shear tensor

$$\bar{N}^{\alpha\beta} = h\bar{\sigma}^{\alpha\beta}$$

and otherwise may apply to this stress system all the notations and formulas of plane stress systems (see pp. 112 through 113).

When the plate becomes unstable, it buckles laterally. Each point of the middle plane undergoes a deflection $u^3 = w$ normal to that plane, which depends only on $x^\alpha$, but not on $x^3 = z$. Points not on the middle plane undergo the same displacement $w$, but also a displacement $u_\alpha$ governed by equation (7.57), which states the assumption of the conservation of normals. The incremental strains $\varepsilon_{\alpha\beta}$ and stresses $\sigma^{\alpha\beta}$ are linear functions of $z$; these stresses produce bending and twisting moments $M^{\alpha\beta}$, but no additions $N^{\alpha\beta}$ to the prebuckling forces $\bar{N}^{\alpha\beta}$. We keep this in mind now when we let $i = 3$ and integrate (10.4) across the plate thickness $h$:

$$\int_{-h/2}^{+h/2} [\sigma^{\beta 3}|_\beta + \sigma^{33}|_3 + (\bar{\sigma}^{\beta\gamma} w|_\gamma)|_\beta + (\bar{\sigma}^{3\gamma} w|_\gamma)|_3 + (\bar{\sigma}^{\beta 3} w|_3)|_\beta + (\bar{\sigma}^{33} w|_3)|_3]\, dz = 0.$$

The second term of the integrand yields

$$\int_{-h/2}^{+h/2} \sigma^{33}|_3\, dz = \int_{-h/2}^{+h/2} \sigma^{33}{}_{,3}\, dz = \left[\sigma^{33}\right]_{-h/2}^{+h/2} = 0$$

because the faces of the plate are free from stress. In the fourth, fifth, and sixth terms we note that $\bar{\sigma}^{3\gamma} = \bar{\sigma}^{\beta 3} = 0$ and $w|_3 = w_{,3} = 0$. In the third term we write

$$(\bar{\sigma}^{\beta\gamma} w|_\gamma)|_\beta = \bar{\sigma}^{\beta\gamma}|_\beta\, w|_\gamma + \bar{\sigma}^{\beta\gamma} w|_{\gamma\beta}$$

and see that the first term on the right vanishes because of (7.10), which applies to the prebuckling stresses $\bar{\sigma}^{\beta\gamma}$ with $\bar{X}^\gamma = 0$. In the integrals that still remain, we note that $w$ and its derivatives do not depend on $z$ and, therefore, may be pulled before the integral sign. We apply (7.62) to find that

$$\int_{-h/2}^{+h/2} (\sigma^{\beta 3}|_\beta + \bar{\sigma}^{\beta\gamma} w|_{\gamma\beta})\, dz = -Q^\beta|_\beta + w|_{\beta\gamma} \bar{N}^{\beta\gamma} = 0$$

or, in changed notation,

$$Q^\alpha|_\alpha = \bar{N}^{\alpha\beta} w|_{\alpha\beta}\,. \tag{10.5}$$

This equation describes the equilibrium between the transverse shear forces $Q^\alpha$ produced by the buckling and the original plane stress system $\bar{N}^{\alpha\beta}$, which, through the curvature $w|_{\alpha\beta}$ of the buckled plate, develops a thrust normal to the middle plane.

To complete our equilibrium statements, we derive a moment equation, multiplying (10.4) by a lever arm $z$ and integrating. We let $i = \alpha$ and at once omit terms containing the vanishing stress components $\bar{\sigma}^{\beta 3}$ and $\bar{\sigma}^{33}$:

$$\int_{-h/2}^{+h/2} [\sigma^{\beta\alpha}|_\beta + \sigma^{3\alpha}|_3 + (\bar{\sigma}^{\beta\gamma} u^\alpha|_\gamma)|_\beta] z\, dz = 0.$$

We apply (7.62c) to the integral of the first term and on the second term we perform the same integration by parts as described on page 128. Since the stress $\sigma^{3\alpha}$ on the surfaces $z = \pm h/2$ is zero, we thus obtain from these integrals

$$-M^{\beta\alpha}|_\beta + Q^\alpha.$$

In the third term of the integrand we use the kinematic relation (7.57):

$$\int_{-h/2}^{+h/2} (\bar{\sigma}^{\beta\gamma} u^\alpha|_\gamma)|_\beta \, z \, dz = \int_{-h/2}^{+h/2} (\bar{\sigma}^{\beta\gamma}|_\beta \, w|_\gamma^\alpha + \bar{\sigma}^{\beta\gamma} w|_{\gamma\beta}^\alpha) z^2 \, dz.$$

Everything in the parentheses is independent of $z$ and can be pulled out of the integral. Furthermore, $\bar{\sigma}^{\beta\gamma}|_\beta = 0$, as we have noted earlier. In the remaining integral we use (7.66) and obtain

$$\bar{\sigma}^{\beta\gamma} w|_{\gamma\beta}^\alpha \frac{K(1-\nu^2)}{E} = \frac{\bar{\sigma}^{\beta\gamma}}{E} (Kw|_{\gamma\beta}^\alpha)(1-\nu^2).$$

The first factor on the right-hand side is of the order of a prebuckling strain $\bar{\varepsilon}_{\beta\gamma}$ and, hence, $\ll 1$, while the second factor is, according to (7.65), of the order of a derivative $M_\gamma^\alpha|_\beta$ of a moment. The entire integral is small in the ratio $\varepsilon : 1$ compared to $M^{\beta\alpha}|_\beta$ and, therefore, negligible. All that remains of our equation is the simple statement

$$M^{\beta\alpha}|_\beta = Q^\alpha, \tag{10.6}$$

identical with equation (7.68) of plate bending when we there delete the external moment $m^\alpha$, which is not present in this case.

When we eliminate $Q^\alpha$ between (10.5) and (10.6), we find the condensed equilibrium statement

$$M^{\beta\alpha}|_{\beta\alpha} = \bar{N}^{\alpha\beta} w|_{\alpha\beta} \,. \tag{10.7}$$

In a plate of isotropic material, the moments $M^{\alpha\beta}$ are connected with the deflection $w$ through the same equation (7.65) as in plate bending, and when we use it to eliminate $M^{\beta\alpha}$ from (10.7), we may make the same simple computation as shown on page 129 and arrive at the differential equation for the buckling deflection of the plate:

$$Kw|_{\alpha\beta}^{\alpha\beta} = -\bar{N}^{\alpha\beta} w|_{\alpha\beta} \,. \tag{10.8}$$

This is a homogeneous, linear equation, which always has the trivial solution $w \equiv 0$. If the prebuckling forces $\bar{N}^{\alpha\beta}$ are proportional to a load parameter $p$, it is possible that, for a certain value of this parameter, a nontrivial solution can be found which satisfies the boundary conditions of a given plate. Such a value of $p$ is called an eigenvalue of the mathematical problem and represents a load for which the plate is in a neutral equilibrium, ready to deflect laterally without provocation.

## References

The theory of elastic stability is conspicuously absent from many books on the theory of elasticity. During the last decades it has developed into a subject of its own, about which a number of books have been written. We mention Timoshenko–Gere [32] as strongly solution-oriented and Ziegler [36] as a study of the fundamental ideas.

CHAPTER 11

# Principal Axes and Invariants

WE HAVE SEEN (on p. 46) how the force transmitted in an arbitrary area element $d\mathbf{A}$ can be described in terms of the nine components $\sigma^{ij}$ of the stress tensor. We may ask whether, among all the area elements passing through a given point, one can find a special one for which the force $\sigma^{ij}\, dA_i$ happens to be normal to the area on which it is acting. In that case the stress would be a simple tension or compression, not accompanied by a shear stress. The normal to such an area element and, hence, the direction of the stress in it, is called a *principal direction* or a *principal axis* of the stress tensor $\sigma^{ij}$.

If the force is to be normal to the area element $d\mathbf{A} = dA_j\, \mathbf{g}^j$, its covariant components $\sigma^i_j\, dA_i$ must be proportional to the covariant components $dA_j$ of that normal. Let the proportionality factor be denoted by $\sigma$; then

$$\sigma^i_j\, dA_i = \sigma\, dA_j \tag{11.1}$$

or

$$(\sigma^i_j - \sigma\delta^i_j)\, dA_i = 0. \tag{11.2}$$

This tensor equation represents three component equations for the three unknowns $dA_i$. These equations are homogeneous and, in general, have only the trivial and meaningless solution $dA_i = 0$. A nontrivial solution exists if and only if the determinant of the nine coefficients vanishes, i.e. if

$$S \equiv \det(\sigma^i_j - \sigma\delta^i_j) = \begin{vmatrix} \sigma^1_1 - \sigma & \sigma^2_1 & \sigma^3_1 \\ \sigma^1_2 & \sigma^2_2 - \sigma & \sigma^3_2 \\ \sigma^1_3 & \sigma^2_3 & \sigma^3_3 - \sigma \end{vmatrix} = 0. \tag{11.3}$$

This is a cubic equation for $\sigma$. It has three roots $\sigma_{(m)}$, where the subscript $(m)$ serves to distinguish the three scalars $\sigma_{(1)}, \sigma_{(2)}, \sigma_{(3)}$, which are not components of a vector. The $\sigma_{(m)}$ are known as the *eigenvalues* of (11.2).

Once these eigenvalues have been calculated, each of them may be introduced separately in (11.2) and will lead to a set $dA_i$, which we shall denote by $dA_{(m)i}$. Since the linear equations (11.2) are homogeneous, they are also satisfied by a multiple $p\, dA_{(m)i}$, where $p$ may be any positive or negative or even a complex number. This means that only the ratios $dA_{(m)1} : dA_{(m)2} : dA_{(m)3}$ are determined, that is, the orientation of the area element, but not its size.

Since (11.3) has real coefficients, at least one of its roots must be real, while the other two may be real or a conjugate complex pair, which would lead to conjugate complex and therefore meaningless solutions $dA_{(m)i}$ of (11.2).

## 11.1. Unsymmetric Tensor

Equations like (11.2) and (11.3) can be written for any second-order tensor and yield three (real or complex) eigenvalues and the corresponding principal axes. We shall now approach this problem in a general form, without reference to the special object of the stress tensor.

We start from a tensor $t^{ij}$ and even drop the symmetry requirement, admitting that

$$t^{ij} \neq t^{ji}, \qquad t^{i}_{\cdot j} \neq t_{j}^{\cdot i}. \tag{11.4}$$

With this tensor we associate an eigenvector $v_i$ through the linear tensor equation

$$(t^{i}_{\cdot j} - T\delta^{i}_{j})v_i = 0, \tag{11.5}$$

which stands for three component equations. Since these are homogeneous, a nonvanishing vector $v_i$ exists only if the coefficient determinant vanishes:

$$\det(t^{i}_{\cdot j} - T\delta^{i}_{j}) = \begin{vmatrix} t^{1}_{\cdot 1} - T & t^{2}_{\cdot 1} & t^{3}_{\cdot 1} \\ t^{1}_{\cdot 2} & t^{2}_{\cdot 2} - T & t^{3}_{\cdot 2} \\ t^{1}_{\cdot 3} & t^{2}_{\cdot 3} & t^{3}_{\cdot 3} - T \end{vmatrix} = 0. \tag{11.6}$$

When we apply the expansion formula (3.23) to the determinant, we obtain the following expression:

$$\begin{aligned} 6 \det(t^{i}_{\cdot j} - T\delta^{i}_{j}) &= (t^{i}_{\cdot l} - T\delta^{i}_{l})(t^{j}_{\cdot m} - T\delta^{j}_{m})(t^{k}_{\cdot n} - T\delta^{k}_{n})\epsilon_{ijk}\,\epsilon^{lmn} \\ &= A - BT + CT^2 - DT^3, \end{aligned}$$

in which the exponents attached to $T$ indicate powers of that scalar, not components. To the coefficients $A \cdots D$ we apply (3.20), (3.21), (3.22) and find

$$
\begin{aligned}
A &= 6 \det t^{i}_{\cdot j}, \\
B &= (\delta^{i}_{l} t^{j}_{\cdot m} t^{k}_{\cdot n} + t^{i}_{\cdot l} \delta^{j}_{m} t^{k}_{\cdot n} + t^{i}_{\cdot l} t^{j}_{\cdot m} \delta^{k}_{n})\epsilon_{ijk}\,\epsilon^{lmn} \\
&= (t^{j}_{\cdot m} t^{k}_{\cdot n}\,\epsilon_{ljk} + t^{i}_{\cdot l} t^{k}_{\cdot n}\,\epsilon_{imk} + t^{i}_{\cdot l} t^{j}_{\cdot m}\,\epsilon_{ijn})\epsilon^{lmn} \\
&= t^{j}_{\cdot m} t^{k}_{\cdot n}(\delta^{m}_{j}\,\delta^{n}_{k} - \delta^{m}_{k}\,\delta^{n}_{j}) + t^{i}_{\cdot l} t^{k}_{\cdot n}(\delta^{l}_{i}\,\delta^{n}_{k} - \delta^{l}_{k}\,\delta^{n}_{i}) + t^{i}_{\cdot l} t^{j}_{\cdot m}(\delta^{l}_{i}\,\delta^{m}_{j} - \delta^{l}_{j}\,\delta^{m}_{i}) \\
&= t^{j}_{\cdot j} t^{k}_{\cdot k} - t^{j}_{\cdot k} t^{k}_{\cdot j} + t^{i}_{\cdot i} t^{k}_{\cdot k} - t^{i}_{\cdot k} t^{k}_{\cdot i} + t^{i}_{\cdot i} t^{j}_{\cdot j} - t^{i}_{\cdot j} t^{j}_{\cdot i} \\
&= 3(t^{i}_{\cdot i} t^{j}_{\cdot j} - t^{i}_{\cdot j} t^{j}_{\cdot i}), \\
C &= (t^{i}_{\cdot l}\,\delta^{j}_{m}\,\delta^{k}_{n} + \delta^{i}_{l}\,t^{j}_{\cdot m}\,\delta^{k}_{n} + \delta^{i}_{l}\,\delta^{j}_{m}\,t^{k}_{\cdot n})\epsilon_{ijk}\,\epsilon^{lmn} \\
&= (t^{i}_{\cdot l}\,\epsilon^{ljk} + t^{j}_{\cdot m}\,\epsilon^{imk} + t^{k}_{\cdot n}\,\epsilon^{ijn})\epsilon_{ijk} \\
&= 2(t^{i}_{\cdot l}\,\delta^{l}_{i} + t^{j}_{\cdot m}\,\delta^{m}_{j} + t^{k}_{\cdot n}\,\delta^{n}_{k}) = 6t^{i}_{\cdot i}, \\
D &= \delta^{i}_{l}\,\delta^{j}_{m}\,\delta^{k}_{n}\,\epsilon_{ijk}\,\epsilon^{lmn} = \epsilon_{ijk}\,\epsilon^{ijk} = 6.
\end{aligned}
$$

The determinantal equation (11.6) thus assumes the form

$$T^3 - t^{i}_{\cdot i} T^2 + \tfrac{1}{2}(t^{i}_{\cdot i} t^{j}_{\cdot j} - t^{i}_{\cdot j} t^{j}_{\cdot i})T - \det t^{i}_{\cdot j} = 0. \tag{11.7}$$

This is a cubic equation for the eigenvalues $T$, called the characteristic equation of the tensor $t^{i}_{\cdot j}$.

The tensor components $t^{i}_{\cdot j}$ are defined in a reference frame $\mathbf{g}_i$. If we switch to another reference frame $\mathbf{g}_{i'}$, the corresponding tensor components $t^{i'}_{\cdot j'}$ will appear in (11.7), but its three solutions $T$ must be the same as before. This is possible only if the coefficients of this equation are the same in both reference frames, that is, if they are invariant against coordinate transformations. We denote these invariants as follows:

$$
\left.
\begin{aligned}
t_I &= t^{i}_{\cdot i}, \\
t_{II} &= t^{i}_{\cdot i} t^{j}_{\cdot j} - t^{i}_{\cdot j} t^{j}_{\cdot i}, \\
t_{III} &= \det t^{i}_{\cdot j}.
\end{aligned}
\right\} \tag{11.8}
$$

They are known as the first, second, and third invariant of the tensor $t^{i}_{\cdot j}$ and are, respectively, a linear, a quadratic, and a cubic function of its components. Any combination of the three is, of course, also invariant. There is little chance to replace $t_I$ by anything simpler or $t_{III}$ by something more obvious, but several combinations of $t_{II}$ and the square of $t_I$ have been suggested to serve as the second invariant, for example

$$t_I^2 - t_{II} = t^{i}_{\cdot j} t^{j}_{\cdot i}, \tag{11.9a}$$

$$2t_I^2 - 3t_{II} = 3t^{i}_{\cdot j} t^{j}_{\cdot i} - t^{i}_{\cdot i} t^{j}_{\cdot j}. \tag{11.9b}$$

The first of these has the advantage of simple appearance while the second one can be expressed in terms of the deviator

$$\bar{t}^{i}_{\cdot j} = t^{i}_{\cdot j} - \tfrac{1}{3} t^{m}_{\cdot m}\,\delta^{i}_{j},$$

which we have introduced for stress and strain tensors in (4.30). The reader may verify that

$$2t_I^2 - 3t_{II} = 3\bar{t}^{\cdot i}_{\cdot j}\,\bar{t}^{\cdot j}_{\cdot i}. \tag{11.10}$$

Let $T_{(m)}$ and $T_{(n)}$ be two distinct roots of (11.6) and let $v_{(m)i}$ and $v_{(n)k}$ be the eigenvectors associated with them through (11.5). We restate this equation in the following form:

$$t^{i}_{\cdot j}\, v_{(m)i} = T_{(m)}\, v_{(m)j} \qquad \text{and} \qquad t^{k}_{\cdot l}\, v_{(n)k} = T_{(n)}\, v_{(n)l}\,.$$

We multiply the second of these equations by $g^{il}$ and process its left-hand side as follows:

$$t^{k}_{\cdot l}\, g^{il} v_{(n)k} = t^{ki} g_{jk}\, v^{j}_{(n)} = t_{j}^{\cdot i} v^{j}_{(n)}\,.$$

This leads to the pair of equations

$$t^{i}_{\cdot j}\, v_{(m)i} = T_{(m)}\, v_{(m)j} \qquad \text{and} \qquad t_{j}^{\cdot i} v^{j}_{(n)} = T_{(n)}\, v^{i}_{(n)}. \tag{11.11a, b}$$

Finally, we multiply (11.11a) by $v^{j}_{(n)}$ and (11.11b) by $v_{(m)i}$:

$$t^{i}_{\cdot j}\, v_{(m)i}\, v^{j}_{(n)} = T_{(m)}\, v_{(m)j}\, v^{j}_{(n)}, \qquad t_{j}^{\cdot i} v_{(m)i}\, v^{j}_{(n)} = T_{(n)}\, v_{(m)i}\, v^{i}_{(n)}\,.$$

The left-hand sides of these equations are equal if and only if $t^{i}_{\cdot j} = t_{j}^{\cdot i}$, that is, if the tensor $t^{ij}$ is symmetric. On the right the factors $T_{(m)}$ and $T_{(n)}$ are definitely different and it follows that, for a symmetric tensor,

$$v_{(m)j}\, v^{j}_{(n)} = 0 \qquad \text{for} \qquad m \neq n, \tag{11.12}$$

which indicates that the eigenvectors associated with two different eigenvalues are orthogonal.† Therefore, if all the eigenvalues are real and different, the principal axes of a symmetric tensor $t^{ij}$ are orthogonal to each other. For unsymmetric tensors no statement of this kind can be made.

Restricting the further study to symmetric tensors, we may investigate the possibility of complex eigenvalues. Let us assume that there is a pair $T_{(1)} = R + iS$ and $T_{(2)} = R - iS$ among the roots of (11.6) and let the corresponding eigenvectors be $v_{(1)j} = u_j + iw_j$ and $v_{(2)j} = u_j - iw_j$ with real quantities $R$, $S$, $u_j$ and $w_j$. Then the orthogonality relation (11.12) requires that

$$(u_j + iw_j)(u^j - iw^j) = u_j u^j + w_j w^j + i(w_j u^j - u_j w^j) = 0.$$

The terms in parentheses are equal and cancel. The remaining terms are each the dot product of a vector by itself, therefore positive and, since their sum cannot be zero, we are forced to admit that there can be no complex eigenvalues $T_{(m)}$.

† If (11.6) has a double root, the linear equations (11.5) have two linearly independent solutions, which may be so chosen that they are orthogonal. Any linear combination of them is also an eigenvector.

We may use the eigenvectors $\mathbf{v}_{(m)} = v^i_{(m)}\mathbf{g}_i$ as base vectors $\mathbf{g}_{m'} = \beta^i_{m'}\mathbf{g}_i$. Their contravariant components $v^i_{(m)}$ are identical with the transformation coefficients $\beta^i_{m'}$. The orthogonality relation (11.12), which may also be written as

$$v^i_{(m)} v^j_{(n)} g_{ij} = 0, \qquad m \neq n,$$

yields

$$\beta^i_{m'} \beta^j_{n'} g_{ij} = g_{m'n'} = 0 \qquad \text{for} \qquad m' \neq n'$$

in agreement with our earlier finding that the reference frame $\mathbf{g}_{m'}$ is orthogonal. Since (11.5) determines only the direction, but not the magnitude of each eigenvector, we may choose the $\mathbf{g}_{m'}$ as unit vectors. If we do so, we have additionally

$$\beta^i_{m'} \beta^j_{n'} g_{ij} = g_{m'n'} = 1 \qquad \text{for} \qquad m' = n'$$

and hence $g_{m'n'} = \delta_{m'n'}$ and the reference frame $\mathbf{g}_{m'}$ is cartesian.

Equation (11.5) may be written in the alternate form

$$t_{ij} v^i_{(m)} = T_{(m)} g_{ij} v^i_{(m)}.$$

Replacing $v^i_{(m)}$ by $\beta^i_{m'}$ and multiplying both sides by $\beta^j_{n'}$ we find that

$$t_{m'n'} = t_{ij}\beta^i_{m'}\beta^j_{n'} = T_{(m)} g_{ij}\beta^i_{m'}\beta^j_{n'} = T_{(m)} g_{m'n'}$$

and hence

$$t_{m'n'} = 0 \qquad \text{for} \qquad m' \neq n'.$$

For the reference frame of the principal axes, the symmetric tensor is thus reduced to diagonal form:

$$\begin{bmatrix} t_{1'1'} & 0 & 0 \\ 0 & t_{2'2'} & 0 \\ 0 & 0 & t_{3'3'} \end{bmatrix} = \begin{bmatrix} T_{(1)} g_{1'1'} & 0 & 0 \\ 0 & T_{(2)} g_{2'2'} & 0 \\ 0 & 0 & T_{(3)} g_{3'3'} \end{bmatrix}$$

The foregoing discussion of principal axes and invariants may also be applied to a two-dimensional tensor $t^{\alpha\beta}$. For such a tensor $t_{II} = 2t_{III}$ and there are only two distinct invariants, $t_I$ and $t_{II}$. Symmetric tensors have a pair of principal axes.

## 11.2. Tensors of Stress and Strain

We may now return to the stress tensor $\sigma^{ij}$, from which we started our investigation of principal axes. Since it is symmetric, there always exists a set of three principal directions, orthogonal to each other. On an area element whose normal

$$d\mathbf{A}_{(m)} = dA_{(m)j}\,\mathbf{g}^j$$

points in a principal direction, according to (11.1) the force

$$dF = \sigma^i_j \, dA_{(m)i} \, g^j = \sigma_{(m)} \, dA_{(m)} \tag{11.13}$$

is acting, which has the direction of the vector $dA_{(m)}$. The *principal stress* $\sigma_{(m)}$ is, therefore, a normal stress, and is one of the roots $\sigma$ of the determinantal equation (11.3), which may be written in the alternate form

$$\det(\sigma^{ij} - \sigma g^{ij}) = 0. \tag{11.14}$$

Since the absolute value of the vector $dA_{(m)}$ is the true area of the element, the stress $\sigma_{(m)}$ is a stress in the common sense of the word, a physical component not in need of a conversion factor.

In a general stress field $\sigma^{ij}(x^k)$ the principal directions change continuously from point to point, and we may draw curves which have everywhere one of these directions for a tangent. These curves are called *stress trajectories*. They form a three-parametric, orthogonal net and may be chosen as coordinate lines of a special, curvilinear coordinate system. The $g_{m'}$ defined on page 176 are its base vectors, but their length is not arbitrary and must be chosen in accordance with the varying mesh width of the net. In section elements normal to one of the principal directions (and tangential to the other two) only a normal stress is transmitted—the principal stress $\sigma_{m'm'}$.

The quantity $s$ defined by (4.29b) is one third of the first invariant $\sigma_I$ as defined by (11.8). The function of the stress components used in the yield condition (4.40) and derived from the distortion energy is the second stress invariant according to (11.9b).

The strain tensor $\varepsilon_{ij}$ is also a symmetric tensor and has three principal axes. Since in a reference frame $g_{m'}$ using the directions of these axes there are no shear strains $\varepsilon_{m'n'}$ $(m' \neq n')$, these axes remain orthogonal during deformation, but the reference frame may, of course, undergo a rigid-body rotation. With reference to these axes, the local deformation consists only of a stretching (possibly negative) in the direction of each of them.

When Hooke's law (4.25) is applied to the reference frame $g_{m'}$ based on the principal directions of strain, it turns out that only the stress components $\sigma^{m'}_{m'}$ with $m' = n'$ are different from zero. This proves that in an isotropic elastic material the principal directions of stress and strain coincide.

In connection with the linear kinematic relation (6.1), we have seen that the strain tensor is the symmetric part of the tensor $u_i|_j$ of the displacement derivatives, which, in general, is not symmetric. We may use this tensor to study the properties of unsymmetric tensors.

The vector $u^j|_i \, dx^i$ describes the displacement of a point $x^i + dx^i$ with respect to a reference frame $g_j$ attached to the point $x^i$ and participating in its translatory motion $u^j$. We may ask whether we can find a line element vector

$dx^i$ such that the relative displacement $u^j|_i\, dx^i$ of its endpoint has the same direction as the vector $dx^i$ itself. In this case there is

$$u^j|_i\, dx^i = \lambda\, dx^j$$

or

$$(u^j|_i - \lambda\delta_i^j)\, dx^i = 0. \tag{11.15}$$

This equation has the same form as (11.5) if we change all subscripts into superscripts and vice versa. The characteristic equation (11.7) here reads

$$\lambda^3 - u^i|_i \lambda^2 + \tfrac{1}{2}(u^i|_i u^j|_j - u^i|_j u^j|_i)\lambda - \det u^i|_j = 0.$$

Now let $x^i$ be a cartesian system with unit vectors $\mathbf{g}^i = \mathbf{i}_i$ as base vectors and consider the plane displacement field

$$u^1 = \alpha x^2, \qquad u^2 = \beta x^1, \qquad u^3 = 0.$$

It leads to the characteristic equation

$$\lambda^3 - \alpha\beta\lambda = 0,$$

which has the roots

$$\lambda = \sqrt{\alpha\beta}, \qquad \lambda = -\sqrt{\alpha\beta}, \qquad \lambda = 0.$$

The corresponding eigenvectors follow from (11.15):

$$\begin{aligned} &dx^2 = \sqrt{\beta/\alpha}\, dx^1, \qquad &&dx^2 = -\sqrt{\beta/\alpha}\, dx^1, \qquad &&dx^1 = dx^2 = 0, \\ &dx^3 = 0, &&dx^3 = 0, &&dx^3 \neq 0. \end{aligned}$$

The third one is normal to the $x^1$, $x^2$ plane and the other two are

$$d\mathbf{r} = (\mathbf{i}_1 \pm \sqrt{\beta/\alpha}\, \mathbf{i}_2)\, dx^1.$$

For $\beta = 0.5\alpha$, they are the vectors $d\mathbf{r}_{(1)}$ and $d\mathbf{r}_{(2)}$ shown in Figure 11.1a and are not orthogonal. The corresponding displacement field is illustrated in Figure 11.1b, which, for several points on a quarter circle, shows the vectors $\mathbf{u} = u^j|_i\, dx^i\, \mathbf{g}_j$. One of them has precisely the direction of the corresponding position vector $d\mathbf{r}$. This is one of the principal directions of the tensor $u^j|_i$. For $\beta = 0$, two of the eigenvectors $d\mathbf{r}$ coincide and $u^j|_i$ has only two principal directions, and for $\beta < 0$, they are complex valued so that there exists only one principal axis. Other unsymmetric tensors show the same behavior.

The tension-and-shear tensor $N^{\alpha\beta}$ and the moment tensor $M^{\alpha\beta}$ of plane plates are defined by (7.62). They are plane, symmetric tensors and each has a pair of orthogonal principal directions. The corresponding tensors for a shell, defined by (9.54) and (9.56), are unsymmetric, but they are almost symmetric, $N^{12} \approx N^{21}$, $M^{12} \approx M^{21}$. Therefore, they always have real

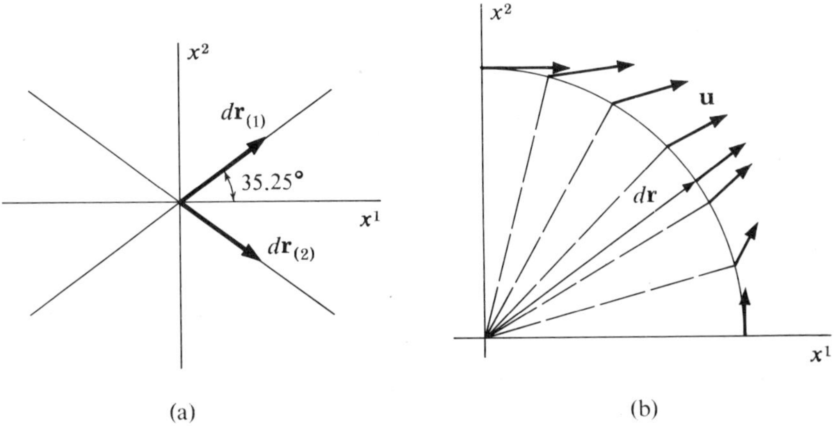


FIGURE 11.1 *Principal axes of the displacement derivative tensor.*

principal directions and these are almost at right angles with each other. On both plates and shells, one can draw force trajectories for $N^{\alpha\beta}$ and moment trajectories for $M^{\alpha\beta}$, which form orthogonal or almost orthogonal nets.

## 11.3. Curvature

Among other second-order tensors which we have encountered in this book, the curvature tensor $b_{\alpha\beta}$ of a curved surface deserves a brief inspection. It has two invariants, the mean curvature $b^{\alpha}_{\alpha}$ and the Gaussian curvature det $b_{\alpha\beta}$, introduced in (8.23) and (8.24), respectively. It is symmetric and has two orthogonal principal directions with base vectors $\mathbf{a}_{\lambda}$, for which $b_{12} = b_{21} = 0$. The components $b_{11}$ and $b_{22}$ are the principal curvatures. As may be read from (8.23) and (8.24), the mean curvature is the sum and the Gaussian curvature is the product of mixed-variant components $b^1_1$ and $b^2_2$ of the principal curvatures. On the surface one may draw curves which are everywhere tangent to one of the directions of principal curvature. They are called *lines of principal curvature* and form an orthogonal net. They have often been used as a coordinate net for the formulation of the basic equations of shell theory (Love [19], Novozhilov [25]). In most cases their use is obvious (surfaces of revolution, cylinders) but in other cases they are unsuitable because the edge of the shell does not coincide with a coordinate line and a skew coordinate net does a better service. It is the strength of the tensor formulation of continuum mechanics that skew coordinate nets can easily be used.

## 11.4. Vectors

In the light of what we have learned about second-order tensors, it is interesting to have a look at vectors. The force vector **P** of (1.19) has a certain direction and we may, if we wish, call this its principal axis. If we transform the components $P^i$ from the reference frame $\mathbf{g}_i$ to a frame $\mathbf{g}_{j'}$, which is so chosen that $\mathbf{g}_{1'}$ has the direction of **P**, then $P^{2'} = P^{3'} = 0$. The base vectors $\mathbf{g}_{2'}$ and $\mathbf{g}_{3'}$ need not be orthogonal to $\mathbf{g}_{1'}$ and have no particular interest, but the principal direction has an obvious physical meaning. For a force vector, it is the direction in which the force is pulling; for a velocity vector it is the direction in which motion takes place.

In a velocity field we may draw curves which have everywhere the local vector **v** as a tangent—the streamlines. In a force field (for example a magnetic force field), we may draw similar field lines tangent to the force vector **P**.

Surprisingly, there exist some vectors to which a physical interpretation of this kind does not apply. The shear force $Q^\alpha$ in a plate is an example. From the definition (7.62a) and from the equilibrium conditions (7.68) and (7.69) it is clear that $Q^\alpha$ satisfies the transformation requirement of a first-order tensor, but the two components $Q^1$ and $Q^2$ represent forces transmitted *in different sections* through the plate and are not the components of one force. However, we may perform a coordinate transformation from the base vectors $\mathbf{g}_\alpha$ to another pair $\mathbf{g}_{\beta'}$. Then, from (1.35c),

$$Q^{\beta'} = Q^\alpha \beta_\alpha^{\beta'}.$$

Now let us assume that $\mathbf{g}_\alpha$ are unit vectors and that $\mathbf{g}_{\beta'}$ are a pair of orthogonal unit vectors as shown in Figure 11.2. Applying (1.33a) to these vectors, one easily verifies that

$$\begin{aligned} \beta_1^{1'} &= \cos\beta, \qquad & \beta_1^{2'} &= -\sin\beta, \\ \beta_2^{1'} &= \sin\gamma, \qquad & \beta_2^{2'} &= \cos\gamma, \end{aligned}$$

whence

$$\begin{aligned} Q^{1'} &= Q^1\beta_1^{1'} + Q^2\beta_2^{1'} = Q^1\cos\beta + Q^2\sin\gamma, \\ Q^{2'} &= Q^1\beta_1^{2'} + Q^2\beta_2^{2'} = -Q^1\sin\beta + Q^2\cos\gamma. \end{aligned}$$

We ask whether we may choose $\beta$ and hence $\gamma = 90° + \beta - \alpha$ so that $Q^{1'}$ becomes a maximum. Differentiating $Q^{1'}$ with respect to $\beta$, we find that

$$\frac{dQ^{1'}}{d\beta} = -Q^1\sin\beta + Q^2\cos\gamma = Q^{2'}.$$

Therefore, for the same orthogonal frame in which $Q^{1'}$ is a maximum, $Q^{2'} = 0$. The vector $\mathbf{g}_{1'}$ points in the direction in which the load is passed on

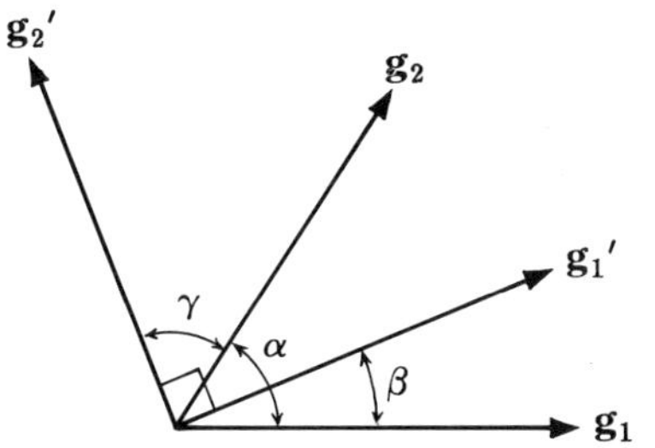


FIGURE 11.2 *Reference frames* $\mathbf{g}_\alpha$ *and* $\mathbf{g}_{\beta'}$.

from one plate element to the next one, while in the direction normal to it no force is transmitted.

Also in this case it is possible to draw field lines following the direction of the vectors $\mathbf{Q} = Q^\alpha \mathbf{g}_\alpha$. This indicates the direction of the force flow which carries the loads on a plate to the supports.

*Problem*

11.1. The torsion stresses $\tau^\alpha = \sigma^{3\alpha}$ in the plane of the cross section are obviously components of a plane vector. Show that the stresses $\tau^\alpha = \sigma^{\alpha 3}$ in longitudinal sections of the bar have the same character as the shear force $Q^\alpha$ in a plate.

CHAPTER 12

# Compilation of Tensor Formulas

THE APPLICATION OF the tensor calculus to the formulation of problems in mechanics requires the skillful use of many formulas. The following list of equations derived in the text will be useful as a ready reference. However, none of the formulas should be applied blindly; the derivation and possible limitation may be found at the proper place.

The formulas have been arranged in logical groups and carry the same numbers as in the text.

## 12.1. Mathematical Formulas

Range convention (see p. 35):

Latin indices $i, j, k, \ldots = 1, 2, 3$

Greek indices $\alpha, \beta, \gamma, \ldots = 1, 2$

Capitals $I, J, \ldots$ and indices in parentheses $(m), (n), \ldots$ are not subjected to the range convention.

Summation convention (see p. 6):

$$a_i b^i = \sum_i a_i b^i, \qquad a_\alpha b^\alpha = \sum_\alpha a_\alpha b^\alpha$$

Kronecker delta:

$$\left.\begin{array}{l} \delta^i_j = 1 \quad \text{if} \quad i = j, \qquad \delta^i_j = 0 \quad \text{if} \quad i \neq j \\ \qquad \delta^i_i = 3, \qquad \delta^\alpha_\alpha = 2 \end{array}\right\} \tag{1.4}$$

Permutation symbol:

$$\begin{aligned} e_{ijk} = e^{ijk} &= +1 \quad \text{if} \quad i, j, k \text{ are cyclic} \\ &= -1 \quad \text{if} \quad i, j, k \text{ are anticyclic} \\ &= \phantom{+}0 \quad \text{if} \quad i, j, k \text{ are acyclic} \end{aligned}$$

(see p. 29)

in two dimensions:

$$e_{11} = e_{22} = 0, \qquad e_{12} = +1, \qquad e_{21} = -1$$
$$e^{\alpha\beta} = e_{\alpha\beta}$$

Permutation tensor:

$$\epsilon_{ijk} = \sqrt{g}e_{ijk}, \qquad \epsilon^{ijk} = e^{ijk}/\sqrt{g} \qquad (3.13), (3.15)$$

$$\epsilon^{ijk}\epsilon_{lmn} = \delta^i_l\,\delta^j_m\,\delta^k_n - \delta^i_l\,\delta^j_n\,\delta^k_m + \delta^i_n\,\delta^j_l\,\delta^k_m - \delta^i_n\,\delta^j_m\,\delta^k_l + \delta^i_m\,\delta^j_n\,\delta^k_l - \delta^i_m\,\delta^j_l\,\delta^k_n \qquad (3.19)$$

$$\epsilon^{ijk}\epsilon_{imn} = \delta^j_m\,\delta^k_n - \delta^j_n\,\delta^k_m \qquad (3.20)$$

$$\epsilon^{ijk}\epsilon_{ijn} = 2\delta^k_n \qquad (3.21)$$

$$\epsilon^{ijk}\epsilon_{ijk} = 6 \qquad (3.22)$$

$$\text{if} \quad t_{ij} = t_{ji}, \qquad \text{then } \epsilon^{ijk}t_{ij} = 0 \qquad (3.16)$$

in two dimensions:

$$\epsilon_{\alpha\beta} = \epsilon_{\alpha\beta 3}, \qquad \epsilon^{\alpha\beta} = \epsilon^{\alpha\beta 3} \qquad (3.26)$$

$$\epsilon^{\alpha\beta}\epsilon_{\gamma\delta} = \delta^\alpha_\gamma\,\delta^\beta_\delta - \delta^\alpha_\delta\,\delta^\beta_\gamma \qquad (3.29a)$$

$$\epsilon^{\alpha\beta}\epsilon_{\gamma\beta} = \delta^\alpha_\gamma, \qquad \epsilon^{\alpha\beta}\epsilon_{\alpha\beta} = 2 \qquad (3.29b, c)$$

Common derivative:

$$\mathbf{u}_{,i} = \frac{\partial \mathbf{u}}{\partial x^i}, \qquad u_{i,j} = \frac{\partial u_i}{\partial x^j} \qquad (2.12)$$

Dot product:

$$\mathbf{u}\cdot\mathbf{v} = u_i v^i = u^i v_i = u_i v_j g^{ij} = u^i v^j g_{ij} \qquad (1.22), (1.13), (1.29)$$

Base vectors:

$$\mathbf{g}_i = \mathbf{r}_{,i}, \qquad \mathbf{g}_i \cdot \mathbf{g}^j = \delta^j_i \qquad (1.18), (1.20)$$

$$\mathbf{v} = v^i\mathbf{g}_i = v_i\mathbf{g}^i \qquad (1.19), (1.21)$$

$$d\mathbf{s} = dx^i\,\mathbf{g}_i \qquad (1.16)$$

Metric tensor:

$$\mathbf{g}_i = g_{ij}\mathbf{g}^j, \qquad \mathbf{g}^i = g^{ij}\mathbf{g}_j \qquad (1.23)$$

$$g_{ij} = \mathbf{g}_i \cdot \mathbf{g}_j, \qquad g^{ij} = \mathbf{g}^i \cdot \mathbf{g}^j \qquad (1.24)$$

$$g^i_j = \delta^i_j \qquad (1.53)$$

$$g_{ik}\,g^{jk} = \delta^j_i \qquad (1.26)$$

$$d\mathbf{s}\cdot d\mathbf{s} = g_{ij}\,dx^i\,dx^j \qquad (1.27)$$

$$v^i g_{ij} = v_j, \qquad v_i g^{ij} = v^j \tag{1.28}$$

$$\left.\begin{matrix} c^{ij} g_{jk} = c^i_{\cdot k}, & c^{ij} g_{ik} g_{jl} = c_{kl} \\ c_{ij} g^{jk} = c_i^{\cdot k}, & c_{ij} g^{ik} g^{jl} = c^{kl} \end{matrix}\right\} \tag{1.46}$$

Coordinate transformation:

$$\beta^i_{j'} = x^i_{\ ,j'}, \qquad \beta^{i'}_j = x^{i'}_{\ ,j} \tag{1.36}$$

$$\beta^j_{i'} \beta^{k'}_j = \delta^{k'}_{i'}, \qquad \beta^{j'}_i \beta^k_{j'} = \delta^k_i \tag{1.32}$$

$$\mathbf{g}_{i'} = \beta^j_{i'} \mathbf{g}_j, \qquad \mathbf{g}^{i'} = \beta^{i'}_j \mathbf{g}^j \tag{1.31}$$

$$\mathbf{g}_i = \beta^{j'}_i \mathbf{g}_{j'}, \qquad \mathbf{g}^i = \beta^i_{j'} \mathbf{g}^{j'} \tag{1.33}$$

$$\left.\begin{matrix} v_{j'} = v_i \beta^i_{j'}, & v_j = v_{i'} \beta^{i'}_j \\ v^{j'} = v^i \beta^{j'}_i, & v^j = v^{i'} \beta^j_{i'} \end{matrix}\right\} \tag{1.35}$$

$$c_{i'j'} = c_{ij} \beta^i_{i'} \beta^j_{j'}, \qquad c_{ij} = c_{i'j'} \beta^{i'}_i \beta^{j'}_j \tag{1.39}$$

$$c^{i'j'} = c^{ij} \beta^{i'}_i \beta^{j'}_j, \qquad c^{ij} = c^{i'j'} \beta^i_{i'} \beta^j_{j'} \tag{1.41}$$

$$c_{i'}^{\ j'} = c_i^{\ j} \beta^i_{i'} \beta^{j'}_j \tag{1.43}$$

Invariants of the tensor $t_{ij}$:

$$\left.\begin{aligned} t_I &= t^i_{.i}, \\ t_{II} &= t^i_{.i} t^j_{.j} - t^i_{.j} t^j_{.i}, \\ t_{III} &= \det t^i_{.j} \end{aligned}\right\} \tag{11.8}$$

Determinants:

$$g = |g_{ij}|, \qquad \frac{1}{g} = |g^{ij}| \qquad (3.9), (3.10)$$

$$\Delta = |\beta^i_{i'}|, \qquad \frac{1}{\Delta} = |\beta^{i'}_i| \tag{3.8}$$

$$a e_{lmn} = a_l^{\ i} a_m^{\ j} a_n^{\ k} e_{ijk} \tag{3.4a}$$

(with $a$ as on p. 29.)

$$a \epsilon_{lmn} = a_l^{\ i} a_m^{\ j} a_n^{\ k} \epsilon_{ijk} \tag{3.17}$$

Cross product:

$$\mathbf{g}_i \times \mathbf{g}_j = \epsilon_{ijk} \mathbf{g}^k, \qquad \mathbf{g}^i \times \mathbf{g}^j = \epsilon^{ijk} \mathbf{g}_k \qquad (3.32), (3.33)$$

Let $\mathbf{u} \times \mathbf{v} = \mathbf{w}$, then

$$w_k = u^i v^j \epsilon_{ijk}, \qquad w^k = u_i v_j \epsilon^{ijk} \qquad (3.35), (3.36)$$

for any vectors **u**, **v**, **w**:

$$\mathbf{u} \times \mathbf{v} \cdot \mathbf{w} = \mathbf{u} \cdot \mathbf{v} \times \mathbf{w} \tag{3.40}$$

$$\mathbf{u} \times \mathbf{v} \cdot \mathbf{w} = u^i v^j w^k \epsilon_{ijk} \tag{3.37}$$

$$\mathbf{g}_1 \times \mathbf{g}_2 \cdot \mathbf{g}_3 = \sqrt{g}, \qquad \mathbf{g}^1 \times \mathbf{g}^2 \cdot \mathbf{g}^3 = 1/\sqrt{g} \qquad (3.41), (3.42)$$

Christoffel symbols:

$$\mathbf{g}_{i,j} = \Gamma_{ijk}\,\mathbf{g}^k = \Gamma^k_{ij}\,\mathbf{g}_k \qquad (5.2)$$

$$\mathbf{g}^i{}_{,j} = -\Gamma^i_{jk}\,\mathbf{g}^k \qquad (5.9)$$

$$\mathbf{g}_{i,j}\cdot\mathbf{g}_k = \Gamma_{ijk}, \qquad \mathbf{g}_{i,j}\cdot\mathbf{g}^k = \Gamma^k_{ij} \qquad (5.3)$$

$$\Gamma^k_{ij}\,g_{kl} = \Gamma_{ijl}, \qquad \Gamma_{ijk}\,g^{kl} = \Gamma^l_{ij} \qquad (5.4)$$

$$\Gamma_{ijk} = \Gamma_{jik}, \qquad \Gamma^k_{ij} = \Gamma^k_{ji} \qquad (5.6)$$

$$g_{ij,k} = \Gamma_{ikj} + \Gamma_{jki} \qquad (5.7)$$

$$2\Gamma_{ijk} = g_{jk,i} + g_{ki,j} - g_{ij,k} \qquad (5.8)$$

Covariant derivative:

$$\mathbf{v}_{,j} = v^i|_j\,\mathbf{g}_i = v_i|_j\,\mathbf{g}^i \qquad (5.10), (5.12)$$

$$d\mathbf{v} = v^i|_j\,\mathbf{g}_i\,dx^j = v_i|_j\,\mathbf{g}^i\,dx^j \qquad (5.14)$$

$$v^i|_j = v^i{}_{,j} + v^k\Gamma^i_{jk}, \qquad v_i|_j = v_{i,j} - v_k\,\Gamma^k_{ij} \qquad (5.11), (5.13)$$

$$v_i|_j\,g^{ik} = v^k|_j, \qquad v^i|_j\,g_{ik} = v_k|_j \qquad (5.16)$$

$$v_i|^k = v_i|_j\,g^{jk}, \qquad v^i|^k = v^i|_j\,g^{jk} \qquad (5.17)$$

for a scalar:

$$p_{,i} = p|_i \qquad (5.19)$$

for a tensor:

$$\left.\begin{aligned} A_{ij}|_k &= A_{ij,k} - A_{lj}\,\Gamma^l_{ik} - A_{il}\,\Gamma^l_{kj} \\ A^i{}_j|_k &= A^i{}_{j,k} + A^l{}_j\,\Gamma^i_{kl} - A^i{}_l\,\Gamma^l_{jk} \\ A_i{}^j|_k &= A_i{}^j{}_{,k} - A_l{}^j\Gamma^l_{ik} + A_i{}^l\Gamma^j_{kl} \\ A^{ij}|_k &= A^{ij}{}_{,k} + A^{lj}\Gamma^i_{kl} + A^{il}\Gamma^j_{kl} \end{aligned}\right\} \qquad (5.23)$$

second derivative:

$$\mathbf{v}_{,jk} = (v_i|_{jk} + v_i|_l\,\Gamma^l_{jk})\mathbf{g}^i = (v^i|_{jk} + v^i|_l\,\Gamma^l_{jk})\mathbf{g}_i \qquad (5.29)$$

$$v_i|_{jk} = (v_{i,j} - v_m\,\Gamma^m_{ij})_{,k} - (v_{l,j} - v_m\,\Gamma^m_{lj})\Gamma^l_{ik} - (v_{i,l} - v_m\,\Gamma^m_{il})\Gamma^l_{kj} \qquad (5.27)$$

$$\left.\begin{aligned} v_i|_{jk} - v_i|_{kj} &= v_m\,R^m{}_{ijk}, \\ R^m{}_{ijk} &= \Gamma^m_{ik,j} - \Gamma^m_{ij,k} + \Gamma^m_{lj}\,\Gamma^l_{ik} - \Gamma^m_{lk}\,\Gamma^l_{ij} \end{aligned}\right. \qquad (5.28)$$

special tensors:

$$g_{ij}|_k = 0, \qquad \epsilon_{ijk}|_l = 0 \qquad (5.24), (5.25)$$

$$\epsilon_{123,i} = \epsilon_{123}\,\Gamma^j_{ji} \qquad (5.26)$$

Vector field operators:

$$\operatorname{grad} p = p|_i\,\mathbf{g}^i \qquad (5.20)$$

$$\operatorname{div} \mathbf{v} = v_i|^i \tag{5.30}$$

$$\operatorname{curl} \mathbf{v} = v_j|_i \epsilon^{ijk} \mathbf{g}_k = v^j|^i \epsilon_{ijk} \mathbf{g}^k \tag{5.31}$$

$$\operatorname{curl} \operatorname{grad} p = 0, \qquad \operatorname{div} \operatorname{curl} \mathbf{v} = 0 \qquad (5.33), (5.34)$$

$$\nabla^2 p = \operatorname{div} \operatorname{grad} p = p|_i^i, \tag{5.32}$$

Integral theorems

divergence theorem (Gauss' theorem) in two dimensions:

$$\int_A \operatorname{div} \mathbf{u} \, dA = \oint_C \mathbf{u} \cdot d\mathbf{n} \tag{5.37'}$$

$$\int_A u^\gamma|_\gamma \epsilon_{\alpha\beta} \, dr^\alpha \, ds^\beta = \oint_C u^\gamma \, dn_\gamma \qquad (5.37)\dagger$$

divergence theorem (Gauss' theorem) in three dimensions:

$$\int_V \operatorname{div} \mathbf{u} \, dV = \int_S \mathbf{u} \cdot d\mathbf{A} \tag{5.40'}$$

$$\int_V u^m|_m \epsilon_{ijk} \, dr^i \, ds^j \, dt^k = \int_S u^m \, dA_m \tag{5.40}$$

circulation theorem (Stokes' theorem):

$$\int_A (\operatorname{curl} \mathbf{u}) \cdot d\mathbf{A} = \oint_C \mathbf{u} \cdot d\mathbf{s} \tag{5.42'}$$

$$\epsilon^{ijk} \epsilon_{mnk} \int_A u_j|_i \, dr^m \, dt^n = \oint_C u_i \, ds^i \qquad (5.42)\dagger$$

Curvature of a surface:

$$b_{\alpha\beta} = \Gamma_{\alpha\beta 3} = -\Gamma_{3\alpha\beta} = -\Gamma_{3\beta\alpha} = \Gamma^3_{\alpha\beta} \tag{8.16}$$

$$b^\alpha_\beta = -\Gamma^\alpha_{3\beta} \tag{8.18}$$

"mean" curvature $= b^\alpha_\alpha$, Gaussian curvature $b = |b^\alpha_\beta|$ (8.23), (8.24)

$$b_{\alpha\beta} = \mathbf{a}_{\alpha,\beta} \cdot \mathbf{a}_3 = -\mathbf{a}_{3,\alpha} \cdot \mathbf{a}_\beta, \qquad (8.13), (8.14)$$

$$d\mathbf{a}_3 = -b_{\alpha\beta} \mathbf{a}^\beta \, dx^\alpha \tag{8.21}$$

$$d\mathbf{a}_3 \cdot d\mathbf{s} = -b_{\alpha\beta} \, dx^\alpha \, dx^\beta \tag{8.22}$$

Covariant derivative on a curved surface:

$$\mathbf{v}_{,\beta} = v_\alpha|_\beta \mathbf{a}^\alpha + v_3|_\beta \mathbf{a}^3 = v^\alpha|_\beta \mathbf{a}_\alpha + v^3|_\beta \mathbf{a}_3 \tag{8.26}$$

$$v_\alpha|_\beta = v_\alpha\|_\beta - v_3 \, b_{\alpha\beta} \tag{8.30}$$

† Note that here in (5.37) and (5.42) the area element $dA$ is a *rectangle* with sides $d\mathbf{r}$, $d\mathbf{s}$ and $d\mathbf{r}$, $d\mathbf{t}$, while in equation (5.42) of the text it is a *triangle*.

$$v_\alpha\|_\beta = v_{\alpha,\beta} - v_\gamma \Gamma^\gamma_{\alpha\beta} \tag{8.29}$$
$$v^\alpha|_\beta = v^\alpha\|_\beta - v^3 b^\alpha_\beta \tag{8.33}$$
$$v^\alpha\|_\beta = v^\alpha{}_{,\beta} + v^\gamma \Gamma^\alpha_{\gamma\beta} \tag{8.32}$$
$$v_\alpha\|_{\beta\gamma} = (v_{\alpha,\beta} - v_\mu \Gamma^\mu_{\alpha\beta})_{,\gamma} - (v_{\lambda,\beta} - v_\mu \Gamma^\mu_{\lambda\beta})\Gamma^\lambda_{\alpha\gamma} - (v_{\alpha,\lambda} - v_\mu \Gamma^\mu_{\alpha\lambda})\Gamma_{\gamma\beta}^{\lambda} \tag{8.48}$$
$$v_\alpha\|_{\beta\gamma} - v_\alpha\|_{\gamma\beta} = v^\delta b \epsilon_{\delta\alpha} \epsilon_{\beta\gamma} \tag{8.49}$$
$$A^\alpha_\beta\|_\gamma = A^\alpha_{\beta,\gamma} + A^\delta_\beta \Gamma^\alpha_{\delta\gamma} - A^\alpha_\delta \Gamma^\delta_{\beta\gamma}$$
$$A^\alpha_\beta\|_{\gamma\delta} - A^\alpha_\beta\|_{\delta\gamma} = A^\zeta_\beta R_{\cdot\zeta\delta\gamma}^{\alpha} + A^\alpha_\zeta R_{\cdot\beta\gamma\delta}^{\xi} \tag{8.50}$$

Gauss–Codazzi equation:

$$b_{\alpha\beta}\|_\gamma = b_{\alpha\gamma}\|_\beta, \qquad b^\alpha_\beta\|_\gamma = b^\alpha_\gamma\|_\beta \tag{8.42}$$

Riemann–Christoffel tensor:

$$R_{\alpha\beta\gamma\delta} = b_{\alpha\gamma} b_{\beta\delta} - b_{\alpha\delta} b_{\beta\gamma} = b\, \epsilon_{\alpha\beta}\, \epsilon_{\gamma\delta} \qquad (8.45), (8.46)$$

Shell geometry

$$\mu^\alpha_\beta = \delta^\alpha_\beta - z b^\alpha_\beta, \qquad \lambda^\alpha_\beta = \delta^\alpha_\beta + z b^\alpha_\beta + z^2 b^\alpha_\gamma b^\gamma_\beta + \cdots \qquad (9.2), (9.5)$$
$$\mu^\alpha_\beta \lambda^\beta_\gamma = \delta^\alpha_\gamma, \qquad \lambda^\alpha_\beta \mu^\beta_\gamma = \delta^\alpha_\gamma, \tag{9.4}$$
$$\mathbf{g}_\alpha = \mu^\beta_\alpha \mathbf{a}_\beta, \qquad \mathbf{g}^\alpha = \lambda^\alpha_\beta \mathbf{a}^\beta \qquad (9.1), (9.3)$$
$$g_{\alpha\beta} = \mu^\gamma_\alpha \mu^\delta_\beta a_{\gamma\delta}, \qquad g^{\alpha\beta} = \lambda^\alpha_\gamma \lambda^\beta_\delta a^{\gamma\delta} \tag{9.6}$$
$$\mu = |\mu^\alpha_\beta| = 1 - z b^\lambda_\lambda + z^2 b = \tfrac{1}{2}\epsilon^{\alpha\beta}\epsilon_{\gamma\delta}\mu^\gamma_\alpha \mu^\delta_\beta \qquad (9.12), (9.14)$$
$$\mu\|_\lambda = \mu \lambda^\alpha_\beta \mu^\beta_\alpha\|_\lambda \tag{9.16}$$
$$\mu_{,3} = -b^\alpha_\beta \lambda^\beta_\alpha \mu \tag{9.17}$$
$$\overline{\Gamma}^\gamma_{\alpha\beta} = \Gamma^\gamma_{\alpha\beta} + \lambda^\gamma_\delta \mu^\delta_\alpha\|_\beta \tag{9.18}$$
$$\overline{\Gamma}^3_{\alpha\beta} = \mu^\delta_\alpha b_{\delta\beta}, \qquad \overline{\Gamma}^\alpha_{3\beta} = -\lambda^\alpha_\delta b^\delta_\beta \qquad (9.19), (9.20)$$
$$\bar{\epsilon}_{\alpha\beta} = \epsilon_{\alpha\beta}\, \mu \tag{9.22}$$

## 12.2. Mechanical Formulas

Moment of a force:

$$\mathbf{M} = \mathbf{r} \times \mathbf{P}, \qquad M_k = r^i P^j \epsilon_{ijk} \qquad (3.45), (3.46)$$

Strain:

$$\gamma_{ij} = 2\varepsilon_{ij} = \hat{g}_{ij} - g_{ij} \qquad (2.5), (2.17)$$
$$\gamma^{ij} = \gamma_{kl} g^{ik} g^{jl} = -\hat{g}^{ij} + g^{ij} \tag{2.10}$$

Stress:

$$dF^j = \sigma^{ij}\, dA_i, \tag{4.6}$$
$$\sigma^{ij} = \sigma^{ji}, \qquad \sigma^i{}_j = \sigma_j{}^i = \sigma^i_j \neq \sigma^j_i \qquad (4.8), (4.9)$$

Hooke's law, anisotropic material:

$$\sigma^{ij} = E^{ijlm}\varepsilon_{lm}, \qquad \varepsilon_{lm} = C_{ijlm}\sigma^{ij} \qquad (4.11), (7.25)$$

$$E^{ijlm} = E^{jilm} = E^{jiml} = E^{lmij} \qquad (4.12), (4.16)$$

Elastic moduli, isotropic material:

$$\begin{aligned} E^{ijlm} &= \frac{E}{2(1+\nu)}\left(\frac{2\nu}{1-2\nu}\, g^{ij}g^{lm} + g^{il}g^{jm} + g^{im}g^{jl}\right) \\ &= \lambda g^{ij}g^{lm} + \mu(g^{il}g^{jm} + g^{im}g^{jl}) \end{aligned} \qquad (4.19)$$

Lamé moduli:

$$\lambda = \frac{E\nu}{(1+\nu)(1-2\nu)} = \frac{2G\nu}{1-2\nu}, \qquad \mu = \frac{E}{2(1+\nu)} = G \qquad (4.17)$$

Hooke's law, isotropic material:

$$\sigma^i_j = \frac{E}{1+\nu}\left(\varepsilon^i_j + \frac{\nu}{1-2\nu}\,\varepsilon^m_m\,\delta^i_j\right) \qquad (4.25)$$

$$E\varepsilon^i_j = (1+\nu)\sigma^i_j - \nu\sigma^m_m\,\delta^i_j \qquad (4.27)$$

$$\sigma^m_m = \frac{E}{1-2\nu}\,\varepsilon^m_m = 3K\varepsilon^m_m \qquad (4.28)$$

$$3e = \varepsilon^m_m, \qquad 3s = \sigma^m_m \qquad (4.29)$$

$$e^i_j = \varepsilon^i_j - e\delta^i_j, \qquad s^i_j = \sigma^i_j - s\delta^i_j \qquad (4.30)$$

$$Ee = (1-2\nu)s, \qquad Ee^i_j = (1+\nu)s^i_j \qquad (4.31)$$

$$s = (3\lambda + 2\mu)e, \qquad s^i_j = 2\mu e^i_j \qquad (4.33)$$

Elastic strain energy density:

$$a = \tfrac{1}{2}\sigma^{ij}\varepsilon_{ij} = \tfrac{1}{2}\sigma^i_j\,\varepsilon^j_i = \tfrac{1}{2}(3se + s^i_j\,e^j_i) \qquad (4.13), (4.37)$$

dilatation energy $= \tfrac{3}{2}se$, distortion energy $= \tfrac{1}{2}s^i_j\,e^j_i$

isotropic material:

$$a = \tfrac{3}{2}(3\lambda + 2\mu)ee + \mu e^i_j\,e^j_i \qquad (4.38)$$

Plasticity

yield condition:

$$s^i_j\,s^j_i = 2k^2, \qquad 3\sigma^i_j\,\sigma^j_i - \sigma^i_i\,\sigma^j_j = 6k^2 \qquad (4.39), (4.40)$$

flow law:

$$\dot{\varepsilon}^i_j = \lambda s^i_j \tag{4.46}$$

Viscous fluid, viscous volume change:

$$s = (3\lambda + 2\mu)\dot{e}, \qquad s^i_j = 2\mu \dot{e}^i_j \tag{4.34}$$

Viscous fluid, elastic volume change:

$$s = 3Ke, \qquad s^i_j = 2\mu \dot{e}^i_j \tag{4.35}$$

Linear viscoelastic material:

$$\sum_k p''_k \frac{\partial^k s}{\partial t^k} = \sum_k q''_k \frac{\partial^k e}{\partial t^k}, \qquad \sum_k p'_k \frac{\partial^k s^i_j}{\partial t^k} = \sum_k q'_k \frac{\partial^k e^i_j}{\partial t^k} \tag{4.36}$$

Kinematic relations

linear:

$$\varepsilon_{ij} = \tfrac{1}{2}(u_i|_j + u_j|_i) \tag{6.1}$$

general:

$$\varepsilon_{ij} = \tfrac{1}{2}(u_i|_j + u_j|_i + u^k|_i u_k|_j) \tag{6.2}$$

compatibility:

$$\varepsilon_{ij}|_{kl}\, \epsilon^{ikm}\, \epsilon^{jln} = 0 \tag{6.4}$$

Equilibrium:

$$\sigma^{ji}|_j + X^i = 0 \tag{6.6}$$

Newton's law:

$$\sigma^{ji}|_j = -X^i + \rho \ddot{u}^i \tag{6.7}$$

Fundamental equation of the theory of elasticity

anisotropic, homogeneous:

$$E^{ijlm}(u_l|_{mj} + u_m|_{lj}) = -2X^i + 2\rho \ddot{u}^i \tag{6.9}$$

isotropic:

$$u^i|^j_j + \frac{1}{1-2\nu} u^j|^i_j = \frac{1}{G}(-X^i + \rho \ddot{u}^i) \tag{6.12}$$

Elastic waves

general:

$$(\lambda + 2\mu)\nabla^2 u_i|^i - \rho \ddot{u}_i|^i = 0 \qquad (6.14), (6.15)$$

$$(\mu \nabla^2 u_i|_k - \rho \ddot{u}_i|_k)\, \epsilon^{ijk} = 0 \qquad (6.16), (6.17)$$

dilatational wave:

$$(\lambda + 2\mu)\nabla^2 u_i - \rho \ddot{u}_i = 0, \qquad u_i|_j \epsilon^{ijk} = 0 \qquad (6.20), (6.18)$$

shear wave:

$$\mu\nabla^2 u_i - \rho \ddot{u}_i = 0, \qquad u^i|_i = 0 \qquad (6.20), (6.21)$$

Incompressible fluid
continuity condition:

$$v^i|_i = 0 \qquad (6.32)$$

Navier–Stokes equation:

$$\rho \dot{v}_i + \rho v_j v_i|^j - \mu v_i|^j_j = X_i - p|_i \qquad (6.37)$$

inviscid flow:

$$v_i = \Phi|_i, \qquad \Phi|^i_i = 0 \qquad (6.43), (6.44)$$

Seepage flow
Darcy's law:

$$v^i = k^{ij}(X_j - p|_j) \qquad (6.47)$$

differential equation of the pressure field, general:

$$(k^{ij} p|_j)|_i = (k^{ij} X_j)|_i \qquad (6.49)$$

differential equation of the pressure field, homogeneous medium:

$$k^{ij} p|_{ij} = k^{ij} X_j|_i \qquad (6.50)$$

gross stress:

$$\sigma^{ij}|_j = -X^i + (p\phi)|^i. \qquad (6.53)$$

Plane strain
Hooke's law:

$$\sigma^\alpha_\beta = \frac{E}{1+\nu}\left(\varepsilon^\alpha_\beta + \frac{\nu}{1-2\nu}\varepsilon^\zeta_\zeta \delta^\alpha_\beta\right) \qquad (7.8)$$

$$\varepsilon^\alpha_\beta = \frac{1+\nu}{E}(\sigma^\alpha_\beta - \nu\sigma^\zeta_\zeta \delta^\alpha_\beta) \qquad (7.9)$$

fundamental equation:

$$u_\alpha|^\beta_\beta = \frac{1}{1-2\nu} u_\beta|^\beta_\alpha + \frac{2(1+\nu)}{E} X_\alpha = 0 \qquad (7.14)$$

Airy stress function
definition:

$$\sigma^{\alpha\beta} = \epsilon^{\alpha\lambda}\epsilon^{\beta\mu}\Phi|_{\lambda\mu} - g^{\alpha\beta}\Omega, \qquad \text{where} \qquad X_\alpha = \Omega|_\alpha; \qquad (7.17)$$

Airy stress function,
differential equation:

$$(1-\nu)\Phi|^{\alpha\beta}_{\alpha\beta} = (1-2\nu)\Omega|^{\alpha}_{\alpha}. \qquad (7.16), (7.23)$$

Plane stress

Hooke's law:

$$\sigma^{\alpha}_{\beta} = \frac{E}{1+\nu}\left(\varepsilon^{\alpha}_{\beta} + \frac{\nu}{1-\nu}\varepsilon^{\zeta}_{\zeta}\,\delta^{\alpha}_{\beta}\right), \qquad (7.29)$$

$$\varepsilon^{\alpha}_{\beta} = \frac{1+\nu}{E}\left(\sigma^{\alpha}_{\beta} - \frac{\nu}{1+\nu}\sigma^{\zeta}_{\zeta}\,\delta^{\alpha}_{\beta}\right) \qquad (7.28)$$

fundamental equation:

$$u_{\alpha}|^{\beta}_{\beta} + \frac{1+\nu}{1-\nu}u_{\beta}|^{\beta}_{\alpha} + \frac{2(1+\nu)}{E}X_{\alpha} = 0. \qquad (7.30)$$

Airy stress function, differential equation:

$$\Phi|^{\alpha\beta}_{\alpha\beta} = (1-\nu)\Omega|^{\alpha}_{\alpha} \qquad (7.31)$$

Torsion

stress function:

$$\tau^{\alpha} = \Phi|_{\gamma}\,\epsilon^{\gamma\alpha}, \qquad \Phi|^{\alpha}_{\alpha} = 2G\theta \qquad (7.51), (7.52)$$

torque:

$$M = 2\iint \Phi\, ds^{\beta}\, dt^{\alpha}\, \epsilon_{\alpha\beta}. \qquad (7.56)$$

Plates

elastic law:

$$M^{\alpha\beta} = -\frac{h^3}{12}\bar{E}^{\alpha\beta\gamma\delta}w|_{\gamma\delta} \qquad (7.63)$$

differential equation:

$$Kw|^{\alpha\beta}_{\alpha\beta} = p^3 - m^{\alpha}|_{\alpha} \qquad (7.72)$$

Shells

strains of the middle surface:

$$\varepsilon_{\alpha\beta} = \tfrac{1}{2}(u_{\alpha}\|_{\beta} + u_{\beta}\|_{\alpha}) - wb_{\alpha\beta}, \qquad (9.38)$$

$$\kappa_{\alpha\beta} = u_{\gamma}\,b^{\gamma}_{\beta}\|_{\alpha} + u_{\gamma}\|_{\alpha}\,b^{\gamma}_{\beta} - u_{\beta}\|_{\gamma}\,b^{\gamma}_{\alpha} + w\|_{\alpha\beta} + wb^{\gamma}_{\alpha}\,b_{\gamma\beta} \qquad (9.45)$$

equilibrium:

$$N^{\beta\alpha}\|_{\beta} + Q^{\beta}b^{\alpha}_{\beta} + p^{\alpha} = 0, \qquad N^{\alpha\beta}b_{\alpha\beta} - Q^{\alpha}\|_{\alpha} + p^3 = 0 \qquad (9.61), (9.63)$$

$$M^{\beta\alpha}\|_{\beta} - Q^{\alpha} + m^{\alpha} = 0, \qquad \epsilon_{\alpha\beta}(N^{\alpha\beta} + b^{\alpha}_{\gamma}M^{\gamma\beta}) = 0 \qquad (9.64), (9.66)$$

reduced equilibrium conditions:

$$\left.\begin{aligned} N^{\beta\alpha}\|_\beta + M^{\beta\gamma}\|_\beta\, b^\alpha_\gamma &= -p^\alpha - m^\beta b^\alpha_\beta \\ N^{\alpha\beta} b_{\alpha\beta} - M^{\alpha\beta}\|_{\alpha\beta} &= -p^3 + m^\alpha\|_\alpha \end{aligned}\right\} \tag{9.67}$$

elastic law:

$$\begin{aligned} N^{\alpha\beta} &= D[(1-\nu)\varepsilon^{\alpha\beta} + \nu\varepsilon^\zeta_\zeta\, a^{\alpha\beta}] \\ &\quad - K\left[\frac{1-\nu}{2}\,[2a^{\beta\delta}b^{\alpha\gamma} + a^{\beta\gamma}b^{\alpha\delta} + a^{\alpha\delta}b^{\beta\gamma} - b^\lambda_\lambda(a^{\alpha\delta}a^{\beta\gamma} + a^{\alpha\gamma}a^{\beta\delta})]\right. \\ &\quad \left. + \nu(a^{\alpha\beta}b^{\gamma\delta} + a^{\gamma\delta}b^{\alpha\beta} - a^{\alpha\beta}a^{\gamma\delta}b^\lambda_\lambda)\right]\kappa_{\gamma\delta} \end{aligned} \tag{9.71}$$

$$\begin{aligned} M^{\alpha\beta} = K[(1-\nu)(b^\alpha_\gamma\,\varepsilon^{\gamma\beta} - b^\lambda_\lambda\,\varepsilon^{\alpha\beta}) + \nu(b^{\alpha\beta} - b^\lambda_\lambda\, a^{\alpha\beta})\varepsilon^\mu_\mu \\ - \tfrac{1}{2}(1-\nu)(\kappa^{\alpha\beta} + \kappa^{\beta\alpha}) + \nu a^{\alpha\beta}\kappa^\lambda_\lambda] \end{aligned} \tag{9.72}$$

Buckling of a plate:

$$K w|^{\alpha\beta}_{\alpha\beta} = -\overline{N}^{\alpha\beta} w|_{\alpha\beta}\,. \tag{10.8}$$

CHAPTER 13

# Formulas for Special Coordinate Systems

IN THE FOLLOWING formulas, **i**, **j**, **k** are a reference frame of unit vectors in the directions of cartesian coordinates $x$, $y$, $z$. All components not explicitly listed and not connected by a symmetry relation to a listed one are zero. Where convenient, the habitual notations for coordinates have been used instead of $x^1$, $x^2$, $x^3$. On the right-hand side of the formulas, letter superscripts indicate contravariant components, but numbers are exponents of powers.

## 13.1. Plane Polar Coordinates

$$r = x^1, \qquad \theta = x^2 \text{ (see Figure 13.1).}$$

$$\mathbf{g}_r = \mathbf{i}\cos\theta + \mathbf{j}\sin\theta, \qquad \mathbf{g}_\theta = -\mathbf{i}r\sin\theta + \mathbf{j}r\cos\theta,$$

$$g_{rr} = 1, \qquad g_{\theta\theta} = r^2, \qquad g^{rr} = 1, \qquad g^{\theta\theta} = 1/r^2,$$

$$\Gamma_{r\theta\theta} = r, \qquad \Gamma_{\theta\theta r} = -r, \qquad \Gamma^r_{\theta\theta} = -r, \qquad \Gamma^\theta_{r\theta} = 1/r.$$

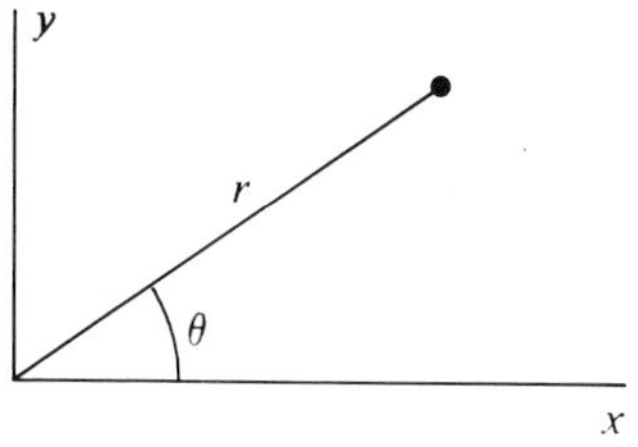


FIGURE 13.1

## 13.2. Plane Elliptic-Hyperbolic Coordinates

Definition: $x = \operatorname{Cosh} \lambda \cos \mu, \qquad y = \operatorname{Sinh} \lambda \sin \mu,$ (see Figure 13.2)

$$\mathbf{g}_\lambda = \mathbf{i} \operatorname{Sinh} \lambda \cos \mu + \mathbf{j} \operatorname{Cosh} \lambda \cos \mu,$$
$$\mathbf{g}_\mu = -\mathbf{i} \operatorname{Cosh} \lambda \sin \mu + \mathbf{j} \operatorname{Sinh} \lambda \cos \mu,$$

$$g_{\lambda\lambda} = g_{\mu\mu} = \operatorname{Cosh}^2 \lambda - \cos^2 \mu, \qquad g^{\lambda\lambda} = g^{\mu\mu} = \frac{1}{\operatorname{Cosh}^2 \lambda - \cos^2 \mu},$$

$$\Gamma_{\lambda\lambda\lambda} = \Gamma_{\lambda\mu\mu} = -\Gamma_{\mu\mu\lambda} = \operatorname{Cosh} \lambda \operatorname{Sinh} \lambda,$$
$$-\Gamma_{\lambda\lambda\mu} = \Gamma_{\lambda\mu\lambda} = \Gamma_{\mu\mu\mu} = \cos \mu \sin \mu,$$

$$\Gamma^\lambda_{\lambda\lambda} = \Gamma^\mu_{\lambda\mu} = -\Gamma^\lambda_{\mu\mu} = \frac{\operatorname{Cosh} \lambda \operatorname{Sinh} \lambda}{\operatorname{Cosh}^2 \lambda - \cos^2 \mu},$$

$$-\Gamma^\mu_{\lambda\lambda} = \Gamma^\lambda_{\lambda\mu} = \Gamma^\mu_{\mu\mu} = \frac{\cos \mu \sin \mu}{\operatorname{Cosh}^2 \lambda - \cos^2 \mu}.$$

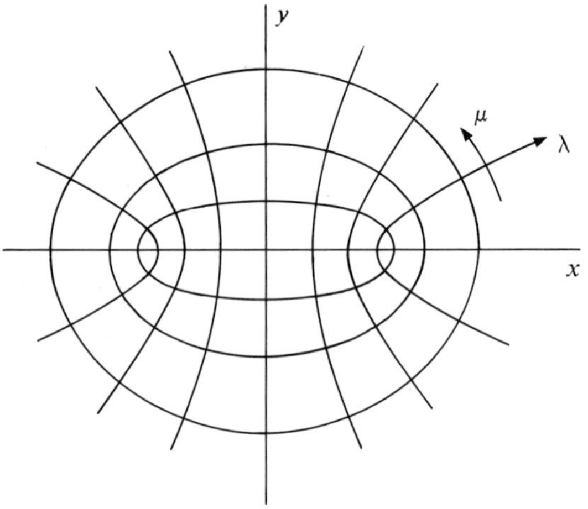


FIGURE 13.2

## 13.3. Plane Bipolar Coordinates

Definition: $x^1 = \psi = \ln\left(\frac{s}{r}\right), \qquad x^2 = \phi$ (see Figure 13.3).

$$\operatorname{Tanh} \psi = \frac{2x}{1 + x^2 + y^2}, \qquad \tan \phi = \frac{2y}{1 - x^2 - y^2},$$

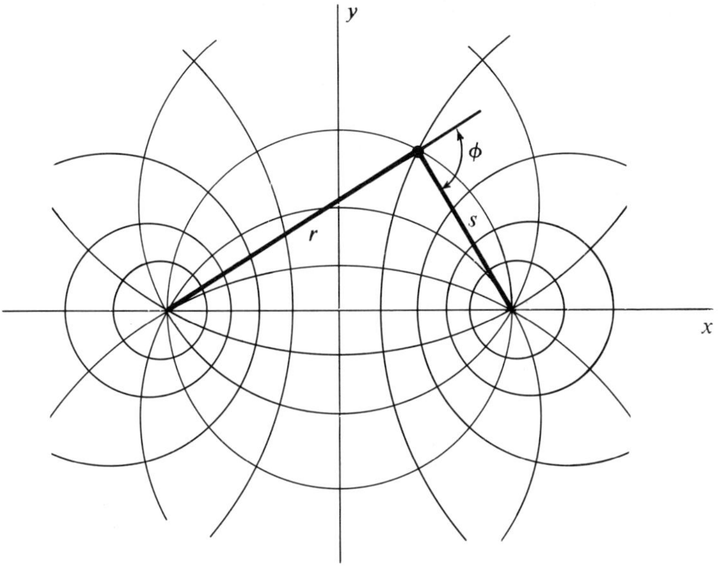


FIGURE 13.3

$$x = \frac{\text{Sinh}\,\psi}{\text{Cosh}\,\psi + \cos\phi}, \qquad y = \frac{\sin\phi}{\text{Cosh}\,\psi + \cos\phi},$$

$$\mathbf{g}_\psi = \mathbf{i}\,\frac{1 + \text{Cosh}\,\psi\cos\phi}{(\text{Cosh}\,\psi + \cos\phi)^2} - \mathbf{j}\,\frac{\text{Sinh}\,\psi\sin\phi}{(\text{Cosh}\,\psi + \cos\phi)^2},$$

$$\mathbf{g}_\phi = \mathbf{i}\,\frac{\text{Sinh}\,\psi\sin\phi}{(\text{Cosh}\,\psi + \cos\phi)^2} + \mathbf{j}\,\frac{1 + \text{Cosh}\,\psi\cos\phi}{(\text{Cosh}\,\psi + \cos\phi^2)},$$

$$g_{\psi\psi} = g_{\phi\phi} = \frac{1}{(\text{Cosh}\,\psi + \cos\phi)^2},$$

$$g^{\psi\psi} = g^{\phi\phi} = (\text{Cosh}\,\psi + \cos\phi)^2,$$

$$\Gamma_{\psi\psi\psi} = \Gamma_{\psi\phi\phi} = -\Gamma_{\phi\phi\psi} = -\frac{\text{Sinh}\,\psi}{(\text{Cosh}\,\psi + \cos\phi)^3},$$

$$\Gamma_{\psi\psi\phi} = -\Gamma_{\psi\phi\psi} = -\Gamma_{\phi\phi\phi} = -\frac{\sin\phi}{(\text{Cosh}\,\psi + \cos\phi)^3},$$

$$\Gamma^{\psi}_{\psi\psi} = \Gamma^{\phi}_{\psi\phi} = -\Gamma^{\psi}_{\phi\phi} = \frac{\text{Sinh}\,\psi}{\text{Cosh}\,\psi + \cos\phi},$$

$$\Gamma^{\psi}_{\psi\phi} = -\Gamma^{\phi}_{\psi\psi} = \Gamma^{\phi}_{\phi\phi} = \frac{\sin\phi}{\operatorname{Cosh}\psi + \cos\phi}.$$

## 13.4. Skew Rectilinear Coordinates

The base vectors are unit vectors: $|\mathbf{g}_1| = |\mathbf{g}_2| = |\mathbf{g}_3| = 1$ (see Figure 13.4).

$$g_{11} = g_{22} = g_{33} = 1,$$
$$g_{12} = \cos\gamma, \qquad g_{23} = \cos\alpha, \qquad g_{13} = \cos\beta,$$
$$g = 1 - \cos^2\alpha - \cos^2\beta - \cos^2\gamma + 2\cos\alpha\cos\beta\cos\gamma,$$

$$g^{11} = \frac{\sin^2\alpha}{g}, \qquad g^{22} = \frac{\sin^2\beta}{g}, \qquad g^{33} = \frac{\sin^2\gamma}{g},$$

$$g^{12} = \frac{\cos\alpha\cos\beta - \cos\gamma}{g}, \qquad g^{23} = \frac{\cos\beta\cos\gamma - \cos\alpha}{g},$$

$$g^{13} = \frac{\cos\alpha\cos\gamma - \cos\beta}{g}.$$

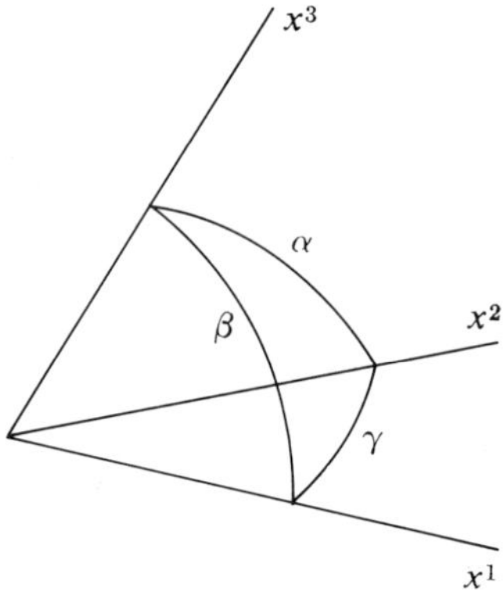


FIGURE 13.4

## 13.5. Cylindrical Coordinates

$x^1 = r$, $x^2 = \theta$, $x^3 = z$ (see Figure 13.5).

All formulas for plane polar coordinates apply and in addition the following:

$$\mathbf{g}_z = \mathbf{k}, \qquad g_{zz} = g^{zz} = 1.$$

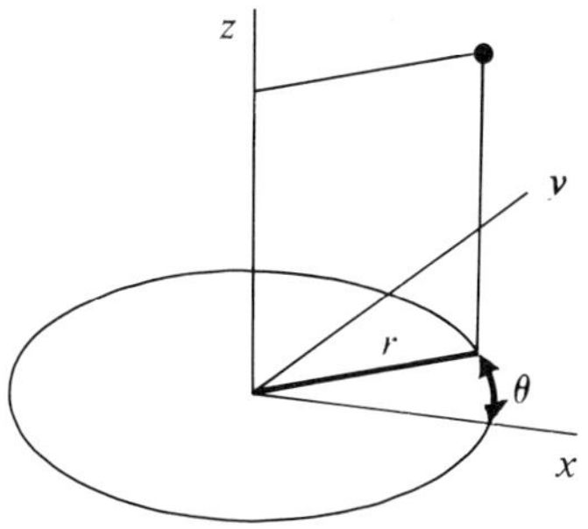


FIGURE 13.5

## 13.6. Spherical Coordinates

$x^1 = r, \qquad x^2 = \theta, \qquad x^3 = \phi$ (see Figure 13.6).

$$
\begin{aligned}
&\mathbf{g}_r = (\mathbf{i}\cos\theta + \mathbf{j}\sin\theta)\cos\phi + \mathbf{k}\sin\phi,\\
&\mathbf{g}_\theta = r(-\mathbf{i}\sin\theta + \mathbf{j}\cos\theta)\cos\phi,\\
&\mathbf{g}_\phi = -r(\mathbf{i}\cos\theta + \mathbf{j}\sin\theta)\sin\phi + \mathbf{k}r\cos\phi,\\
&g_{rr} = 1, \qquad g_{\theta\theta} = (r\cos\phi)^2, \qquad g_{\phi\phi} = r^2,\\
&g^{rr} = 1, \qquad g^{\theta\theta} = (r\cos\phi)^{-2}, \qquad g^{\phi\phi} = r^{-2},\\
&\Gamma_{r\phi\phi} = -\Gamma_{\phi\phi r} = r, \qquad \Gamma_{r\theta\theta} = -\Gamma_{\theta\theta r} = r\cos^2\phi,\\
&\Gamma_{\theta\theta\phi} = -\Gamma_{\theta\phi\theta} = r^2\cos\phi\sin\phi,\\
&\Gamma^r_{\theta\theta} = -r\cos^2\phi, \qquad \Gamma^\phi_{\theta\theta} = \cos\phi\sin,\phi \qquad \Gamma^r_{\phi\phi} = -r,\\
&\Gamma^\theta_{r\theta} = \Gamma^\phi_{r\phi} = 1/r, \qquad \Gamma^\theta_{\theta\phi} = -\tan\phi.
\end{aligned}
$$

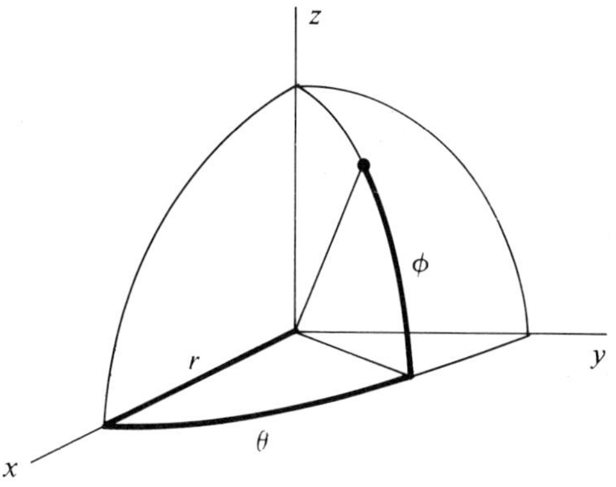


FIGURE 13.6

## 13.7. Skew Circular Cone

Coordinates $x^\alpha$ on the conical surface are $x^1 = z$, $x^2 = \theta$ (see Figure 13.7). Lines $z = \text{const}$ are horizontal circles; lines $\theta = \text{const}$ are straight generators.

$$\mathbf{a}_z = \frac{1}{h}\,[\mathbf{i}(c + b\cos\theta) + \mathbf{j}b\sin\theta + \mathbf{k}h],$$

$$\mathbf{a}_\theta = \frac{1}{h}\,(-\mathbf{i}bz\sin\theta + \mathbf{j}bz\cos\theta),$$

$$a_{zz} = \frac{1}{h^2}\,(h^2 + b^2 + c^2 + 2bc\cos\theta),$$

$$a_{z\theta} = -\frac{bcz}{h^2}\sin\theta, \qquad a_{\theta\theta} = \frac{b^2z^2}{h^2},$$

$$a^{zz} = \frac{h^2}{h^2 + (b + c\cos\theta)^2}, \qquad a^{z\theta} = \frac{h^2}{bz}\,\frac{c\sin\theta}{h^2 + (b + c\cos\theta)^2},$$

$$a^{\theta\theta} = \frac{h^2}{b^2z^2}\,\frac{h^2 + b^2 + c^2 + 2bc\cos\theta}{h^2 + (b + c\cos\theta)^2},$$

$$b_{\theta\theta} = -\frac{bz}{[h^2 + (b + c\cos\theta)^2]^{1/2}},$$

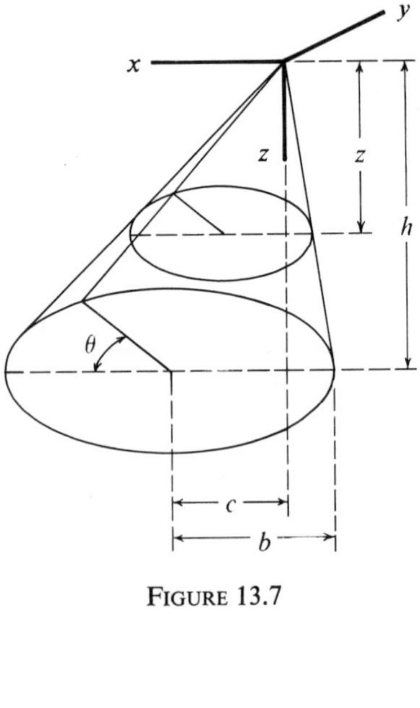


FIGURE 13.7

$$b^z_\theta = -\frac{h^2 c \sin\theta}{[h^2 + (b + c\cos\theta)^2]^{3/2}},$$

$$b^\theta_\theta = -\frac{h}{bz}\frac{h^2 + b^2 + c^2 + 2bc\cos\theta}{[h^2 + (b + c\cos\theta)^2]^{3/2}},$$

$$b^{zz} = -\frac{h^4c^2}{bz}\frac{\sin^2\theta}{[h^2 + (b + c\cos\theta)^2]^{5/2}},$$

$$b^{z\theta} = -\frac{h^4 c\sin\theta}{b^2z^2}\frac{h^2 + b^2 + c^2 + 2bc\cos\theta}{[h^2 + (b + c\cos\theta)^2]^{5/2}},$$

$$b^{\theta\theta} = -\frac{h^4}{b^3z^3}\frac{(h^2 + b^2 + c^2 + 2bc\cos\theta)^2}{[h^2 + (b + c\cos\theta)^2]^{5/2}},$$

$$\Gamma_{z\theta z} = -\frac{bc}{h^2}\sin\theta, \qquad \Gamma_{z\theta\theta} = -\Gamma_{\theta\theta z} = \frac{b^2 z}{h^2},$$

$$\Gamma^\theta_{z\theta} = \frac{1}{z}, \qquad \Gamma^z_{\theta\theta} = -\frac{b^2 z}{h^2 + (b + c\cos\theta)^2},$$

$$\Gamma^\theta_{\theta\theta} = -\frac{bc\sin\theta}{h^2 + (b + c\cos\theta)^2},$$

$$R_{\delta\alpha\beta\gamma} = 0.$$

## 13.8. Right Circular Cone

Coordinates $x^\alpha$ are defined in the same way as for the skew cone. (Other definitions, see Figure 13.8.)

$$\mathbf{a}_z = (\mathbf{i}\cos\theta + \mathbf{j}\sin\theta)\cot\alpha + \mathbf{k},$$
$$\mathbf{a}_\theta = z(-\mathbf{i}\sin\theta + \mathbf{j}\cos\theta)\cot\alpha,$$

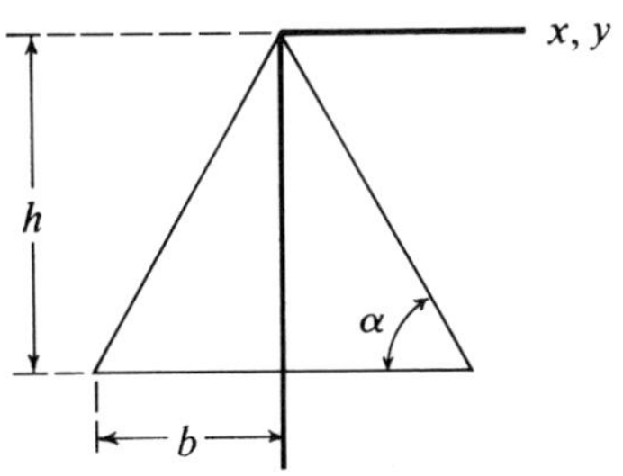


FIGURE 13.8

$$a_{zz} = \frac{1}{\sin^2 \alpha}, \qquad a_{\theta\theta} = z^2 \cot^2 \alpha,$$

$$a^{zz} = \sin^2 \alpha, \qquad a^{\theta\theta} = \frac{1}{z^2} \tan^2 \alpha,$$

$$b_{\theta\theta} = -z \cos \alpha, \qquad b_\theta^\theta = -\frac{1}{z} \sin \alpha \tan \alpha,$$

$$b^{\theta\theta} = -\frac{1}{z^3} \sin \alpha \tan^3 \alpha,$$

$$\Gamma_{z\theta\theta} = -\Gamma_{\theta\theta z} = z \cot^2 \alpha,$$
$$\Gamma_{z\theta}^\theta = z^{-1}, \qquad \Gamma_{\theta\theta}^z = -z \cos^2 \alpha.$$

## 13.9. Hyperbolic Paraboloid

Equation of the surface: $z = xy/c$ (see Figure 13.9). Coordinates $x^\alpha$ on the surface are $x^1 = x$, $x^2 = y$.

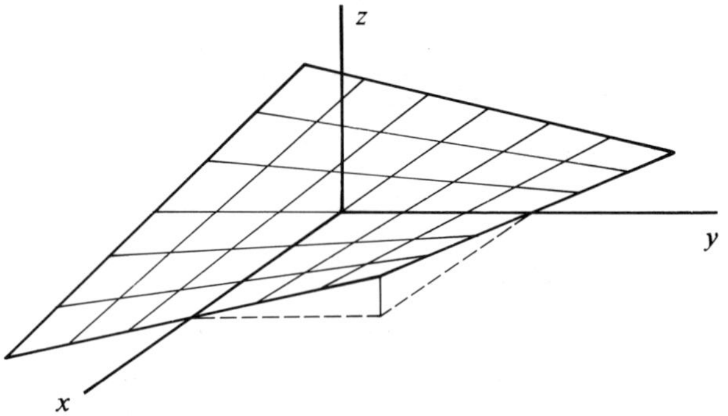


FIGURE 13.9

$$\mathbf{a}_x = \mathbf{i} + \mathbf{k}\frac{y}{c}, \qquad \mathbf{a}_y = \mathbf{j} + \mathbf{k}\frac{x}{c},$$

$$a_{xx} = 1 + \left(\frac{y}{c}\right)^2, \qquad a_{xy} = \frac{xy}{c^2}, \qquad a_{yy} = 1 + \left(\frac{x}{c}\right)^2,$$

$$a^{xx} = \frac{c^2 + x^2}{c^2 + x^2 + y^2}, \qquad a^{xy} = -\frac{xy}{c^2 + x^2 + y^2}, \qquad a^{yy} = \frac{c^2 + y^2}{c^2 + x^2 + y^2},$$

$$b_{xy} = \frac{1}{(c^2 + x^2 + y^2)^{1/2}},$$

$$b_x^x = b_y^y = -\frac{xy}{(c^2 + x^2 + y^2)^{3/2}},$$

$$b_y^x = \frac{c^2 + x^2}{(c^2 + x^2 + y^2)^{3/2}}, \qquad b_x^y = \frac{c^2 + y^2}{(c^2 + x^2 + y^2)^{3/2}},$$

$$b^{xx} = -\frac{2xy(c^2 + x^2)}{(c^2 + x^2 + y^2)^{5/2}},$$

$$b^{xy} = \frac{c^4 + c^2(x^2 + y^2) + 2x^2y^2}{(c^2 + x^2 + y^2)^{5/2}},$$

$$b^{yy} = -\frac{2xy(c^2 + y^2)}{(c^2 + x^2 + y^2)^{5/2}},$$

$$\Gamma_{xyx} = \frac{y}{c^2}, \qquad \Gamma_{xyy} = \frac{x}{c^2},$$

$$\Gamma_{xy}^x = \frac{y}{c^2 + x^2 + y^2}, \qquad \Gamma_{xy}^y = \frac{x}{c^2 + x^2 + y^2},$$

$$R_{xyxy} = R_{yxyx} = -R_{xyyx} = -R_{yxxy} = -\frac{1}{c^2 + x^2 + y^2}.$$

# Bibliography


[1] R. Aris, *Vectors, Tensors, and the Basic Equations of Fluid Mechanics* (Englewood Cliffs, N.J.: Prentice-Hall, 1962).

[2] H. D. Block, *Introduction to Tensor Analysis* (Columbus, Ohio: Merrill, 1962).

[3] L. Brillouin, *Les Tenseurs en Mécanique et en Elasticité* (Paris: Masson, 1946).

[4] N. Coburn, *Vector and Tensor Analysis* (New York: Macmillan, 1955).

[5] A. Duschek and A. Hochrainer, *Grundzüge der Tensorrechnung in analytischer Darstellung* (Wien: Springer, 1946, 1961, 1955), Vol. 1, 2, 3.

[6] C. Eringen, *Mechanics of Continua* (New York: Wiley, 1967).

[7] J. Ferrandon, "Les lois de l'écoulement de filtration," *Génie Civil, 125* (1948), 24–28.

[8] W. Flügge, "Die Stabilität der Kreiszylinderschale," *Ing.-Arch., 3* (1932), 463–506.

[9] W. Flügge, "Die Spannungen in durchströmten porösen Körpern," Abstr. of Papers, *4th Internat. Congr. Appl. Mech.*, Cambridge, England, 1934, pp. 30–31.

[10] W. Flügge, *Stresses in Shells* (Berlin: Springer, 1960).

[11] W. Flügge, *Statik und Dynamik der Schalen* (3rd ed.) (Berlin, Springer, 1962).

[12] K. Girkmann, *Flächentragwerke* (5th ed.) (Wien: Springer, 1959).

[13] A. E. Green and W. Zerna, *Theoretical Elasticity* (Oxford: Clarendon, 1954).

[14] G. A. Hawkins, *Multilinear Analysis for Students in Engineering and Science* (New York: Wiley, 1963).

[15] W. Jaunzemis, *Continuum Mechanics* (New York: Macmillan, 1967).

[16] H. Lamb, *Hydrodynamics* (6th ed.) (New York: Dover, 1945).

[17] H. Lass, *Elements of Pure and Applied Mathematics* (New York: McGraw-Hill, 1957).

[18] T. Levi-Civita, *The Absolute Differential Calculus* (New York: Hafner, 1926).

[19] A. E. H. Love, *A Treatise on the Mathematical Theory of Elasticity* (4th ed.) (New York: Dover, 1944).
[20] E. Madelung, *Die mathematischen Hilfsmittel des Physikers* (7th ed.) (Berlin: Springer, 1964).
[21] M. B. Marlowe, "Some New Developments in the Foundations of Shell Theory" (Ph.D. thesis, Stanford, 1968).
[22] L. M. Milne-Thomson, *Theoretical Hydrodynamics* (3rd ed.) (New York: Macmillan, 1935).
[23] L. M. Milne-Thomson, *Antiplane Elastic Systems* (New York: Academic Press and Berlin: Springer, 1962).
[24] P. M. Naghdi, "Foundations of elastic shell theory," *Progress in Solid Mech.*, *4* (1963), 1–90.
[25] V. V. Novozhilov, *The Theory of Thin Shells*, transl. by P. G. Lowe (Groningen: Noordhoff, 1959).
[26] W. Prager, *Introduction to Mechanics of Continua* (Boston, Mass.: Ginn, 1961).
[27] A. E. Scheidegger, *The Physics of Flow through Porous Media* (New York: Macmillan, 1960).
[28] A. Schild, "Tensor Analysis," in W. Flügge (ed.), *Handbook of Engineering Mechanics* (New York: McGraw-Hill, 1962).
[29] I. S. Sokolnikoff, *Mathematical Theory of Elasticity* (2nd ed.) (New York: McGraw-Hill, 1956).
[30] I. S. Sokolnikoff, *Tensor Analysis* (2nd ed.) (New York: Wiley, 1964).
[31] J. L. Synge and A. Schild, *Tensor Calculus* (Toronto: Univ. of Toronto Press, 1949).
[32] S. P. Timoshenko and J. M. Gere, *Theory of Elastic Stability* (2nd ed.) (New York: McGraw-Hill, 1961).
[33] S. P. Timoshenko and J. N. Goodier, *Theory of Elasticity* (3rd ed.) (New York: McGraw-Hill, 1970).
[34] S. P. Timoshenko and S. Woinowsky-Krieger, *Theory of Plates and Shells* (2nd ed.) (New York: McGraw-Hill, 1959).
[35] A. P. Wills, *Vector Analysis* (New York: Prentice-Hall, 1931).
[36] H. Ziegler, *Principles of Structural Stability* (Waltham, Mass.; Blaisdell, 1968).

# Index

Offset printing: fotokop wilhelm weihert, Darmstadt
Binding: Konrad Triltsch, Graphischer Betrieb, Würzburg